Making Your Small Farm Profitable

Making Your Small Farm Profitable

RON MACHER

Foreword by Howard W. "Bud" Kerr, Jr.

STOREY BOOKS

*The mission of Storey Communications is to serve our
customers by publishing practical information that encourages
personal independence in harmony with the environment.*

Edited by Deborah Burns and Marie Salter
Cover design by Meredith Maker
Cover photograph © by Larry Lefever/Grant Heilman Photography, Inc.
Text design and production by Erin Lincourt

Photographs by Larry Lefever/Grant Heilman Photography, Inc., except those
on page 232 by Nick DeCandia; 98 and 150 by Jane Grushow/Grant Heilman
Photography, Inc.; 50, 74, 184 by Grant Heilman/Grant Heilman Photography, Inc.;
42 by Joel Sartore/Grant Heilman Photography, Inc.; 16 by Kristi Ann Gilman-Miller.

Line drawings by Chuck Galey, except those on pages 12, 56, 82, 161 (bottom)
by Cathy Baker; 161 (top) by Brigita Fuhrmann; 161 (middle) by Millie Holderread;
54 and 55 by Alison Kolesar; 31, 39, 78, 83, 95, 125, 130, 136, 145, 148,
and 187 by Elayne Sears; 162 by Becky Turner.

Indexed by Nan Badgett/Word•a•bil•i•ty

Printed in Canada by Transcontinental Printing
10 9 8 7 6 5 4 3 2 1

Library of Congress Cataloging-in-Publication Data
Macher, Ron.
    Making your small farm profitable / Ron Macher.
        p.   cm.
    Includes bibliographical references and index
    ISBN 1-58017-161-3
    1. Farm management. 2. New agricultural enterprises. I. Title.
S561. M24   1999
630'.68—dc21
                                                                99-16219
                                                                    CIP

Dedication

*To my wife, Joanne,
and to my children, Jean and Jeff*

Contents

Foreword

The agricultural industry is as old as America.

Generations before Columbus discovered the New World, natives of the Western Hemisphere grew maize, squash, and root crops. Our ancestors tilled the soil for subsistence and later embraced farming as a vocation. Over the years, because of scientific breakthroughs, new technology, and improved systems, the number of people employed in farming has declined; still, the business of farming remains vital to our well-being as a people and a nation.

Most of America's nearly two million farms are considered "small," with seven out of ten grossing less than $50,000 a year. Despite their preponderance, operators of small farms have often felt neglected by our national farm programs. Sources of advice for farmers starting out have about dried up, with agricultural county agents admitting that they have time to service only full-time farmers — a group whose numbers are declining. The state departments of agriculture have marketing advice aplenty but are of little help to newcomers asking questions about credit, cropping recommendations, and cultural information. Needed are individual human beings whose hands are on the rural pulse and who have lots of information in their heads, but it remains to be seen who will train them or who will pay them. People have been spoiled for more than a century, recipients of free advice from the government, but that advice is gone now and won't be coming back.

A big establishment and greater sales volume do not guarantee a corresponding increase in profits. Likewise, buying land and calling that land a farm cannot ensure that your investment will be profitable. The days of starting out on a few dollars are over, and farming is now complex. People in the business are specialists, and many even have training in business theory. Such is farming in 1999, and so it will be in the years to come.

Today, as well as tomorrow, the most important piece of farm equipment is knowledge. Understanding complex situations or agribusiness production and marketing problems will be paramount to staying in business year after year. In this era of high technology, vacillating consumer wants and tastes, and shifting market conditions, farm managers and agricultural entrepreneurs are seeking help to cope with these and other situations. In the past — indeed, too frequently — people would "shut the gate after the horse was out of the barn" and then go look for the horse. Now, the preferred method is to have a plan of action in place before going to the barn — knowledge is the key.

It starts with you. You must decide what is good for you, your family, and the farm business before purchasing resources or even planting a seed. Anyone can own a farm and call themselves a farmer, but to become profitable you must acquire and apply business skills.

Making Your Small Farm Profitable is your road map for planning a successful journey into the vast and diverse landscape of agriculture in the years ahead. Ron Macher is a veteran farmer, a true friend of mine and yours, and a huge advocate of small farms. Ron has "been there and done that" for agriculture for many decades. He writes from the heart for the sole purpose of instructing and guiding novice entrepreneurs, wanna-bes, and tried and true "dirt under the fingernails" farm people. Ron, like me, is close to the earth — we are both proven small farm operators. We have been close friends and colleagues championing America's small farm community for many years. (Ron was in the private sector, while I served the public via the U.S. Department of Agriculture.)

In the early 1980s we witnessed the tragic loss of many family farms. Ron did not stand idly by as this occurred; rather, he launched a new publication aimed at Midwest farm families. The fledgling farm magazine *Missouri Farm* grew in popularity, and later the magazine's name was changed to *Small Farm Today* — now the industry bellwether.

Over the years, with the advent of new crops and technology, Ron found that his working hours were increasingly devoted to being an executive, not a farmer. This role brought a new kind of obligation, and Ron soon felt the kind of responsibility he has today; he got a glimpse of the complicated future he and others would face in industrialized agriculture. Already, he was greatly disturbed by the failure of government, both federal and state, to provide the information that small farm operators required. He took it upon himself to meet that need and began editing and publishing his magazine, now in circulation for more than 15 years.

Making Your Small Farm Profitable is the outgrowth of Ron's vast experience in farming and his knowledge of journalism. Over the years, he learned how to share this experience and delegate the details of his business to others. Written for people who want to be profitable regardless of their farm status, novice or veteran, full-time or part-time, this book is your guide from the earliest stages of farming, when all things and topics are nebulous, to a genuine high point in farming — when you make a bank deposit, confirming that your small farm *is* profitable.

Howard W. "Bud" Kerr, Jr.

Former Director, Office of Small-Scale Agriculture,

U.S. Department of Agriculture

Preface

When you are disking a field to plant corn, the sun is shining, and the earth smells fresh, you are probably not thinking about whether that process will make you money. Farmers invariably find production agriculture more fun than the business side of farming, which involves heady subjects like marketing, sales, and cost evaluations. Still, to become a successful agripreneur, you'll need to learn to enjoy the business side of farming as well. It is my sincere hope that this book will help beginning and established farmers make their farms profitable and their livelihoods satisfying, moving us forward to a more consistently sustainable agriculture and ultimately fostering sustainable communities.

What Is Agripreneurship?

Agripreneurship is the profitable marriage of agriculture and entrepreneurship — more plainly, turning your farm into a business. Most farmers regard agriculture as a combination of philosophy and lifestyle, so in its broadest sense agripreneurship binds together philosophy, lifestyle, and business, yielding ideals that give you purpose and goals to strive for. Agripreneurship is a mental attitude that can give you the strength and motivation to break from tradition.

About Sustainable Agriculture

Sustainable agriculture is a site-specific, whole-farm approach to agriculture. Land, people, goals, capital, crops, and livestock are managed to yield the best possible results on the farm. Not tied to any one model of agriculture, sustainable agriculture strives to reduce costs and increase the efficiency of the family farm.

By reducing inputs, sustainable agriculture encourages conservation and multiple uses of resources. It promotes diversity, using multiple species and natural methods to recycle matter and nutrients to maintain the land's productivity, now and in the future. Sustainable agriculture encourages local food production, providing food to society at a reasonable cost while supporting the farmer with an acceptable level of income. Finally, sustainable agriculture fosters a diverse and sustainable farming community as well as a sustainable society, drawing together farmers, lenders, consumers, and institutions in cooperative partnerships.

A Sustainable Community

When considered strictly from an agricultural standpoint, we quickly recognize that society comprises four major groups: farmers, lenders, consumers, and institutions.

❖ Farmers are those who provide food and fiber to our nation. Their farms provide a reserve of natural habitat for wildlife, supply oxygen, and act as filters for our watersheds. Farmers instill in their children a strong work ethic and a good set of values.

❖ Lenders are important suppliers of start-up capital for companies of all sorts. In agriculture, lenders supply monies to agripreneurs and processing facilities, allowing farmers to sell their products directly at retail prices.

❖ Consumers are those who buy the products that individual farmers and institutions produce. Consumers provide farmers with the income they need to continue their operations.

❖ Institutions include universities, government agencies, and businesses. Institutions create consumer goods, add value to products, and perform research.

These four groups are the cornerstones of our society, the community of America. To be a sustainable community, the community requires sustainable agriculture. Respect and cooperation among these groups promotes just that.

Why Do You Need This Book?

Making Your Small Farm Profitable provides a blueprint for farmers. In one volume, I try to tie together the whole picture of farming, show new farmers what they can do and how to choose where to begin, and share my keys to success. I have been a farmer for 34 years and have made a lot of mistakes in that time; some people call this "experience." My fifteen years as a publisher have exposed me to thousands of farm failures and farm success stories. This book should keep you from making some of the mistakes I have made and seen others make — mistakes that can potentially cost thousands of dollars. All of this useful knowledge will be yours with a modest investment of reading time and by practicing the thinking process detailed here.

About This Book

Back in the first rural crisis in 1984, I started a magazine for small farmers (now *Small Farm Today*), because information they needed was not readily available. Recently, I decided to write a book about farming, because the information for new farmers and farmers trying to survive in this age of industrial agriculture is not readily available. There are many good books about individual species of livestock, many useful gardening books, many great books about the reasons for and philosophy of farming, but few books that tie all these aspects together or that help the reader choose where to start.

This book takes a how-to approach — it does not specifically detail how to raise, say, cattle or pumpkins, but rather how to farm successfully and profitably. It is a whole-farm planning approach that ties together outside and on-farm resources with your personal, family, and farm goals. It's about the principles that make your thinking process work like a well-oiled machine.

Throughout the book, you'll find called out in the text pearls of wisdom, guiding principles that, if followed, will help make your farming life more manageable and productive. We begin with a discussion of the basics, information and techniques that will assist you to get started or that will help improve your existing operation. Many of these are things I wish I had known years ago when I was starting out.

Later chapters discuss principles of good farming, resources (location, soil, water, and climate among them), and basic farming methods to help you determine what and how you want to farm.

Then we help you refine your goals, following that up with a discussion of marketing, enterprise choices, machinery, and management. In the final chapters, we tie it all together and send you out as an agripreneur.

You *can* make money on your farm. You *can* enjoy it. Read on to find out how.

Acknowledgments

My special thanks to my wife, Joanne, for the use of half her kitchen table for a year, and for her tolerance of the many piles of books and papers that covered it. I also thank Deborah Burns and Marie Salter of Storey Books, for their good ideas and professionalism. To the readership of *Small Farm Today* magazine — you asked for a book on making a small farm pay — here it is.

Finally, I am grateful for the help and assistance of Paul Berg, editor at *Small Farm Today*. Writers draw strength and inspiration from countless sources, and many people influenced what I have written. In particular, Paul was the prodder who helped me create the most useful book I could. His editing skills and suggestions enabled me to say what I wanted in the best possible manner. My thanks to Paul for all the hours and effort he expended in helping me get this book completed.

PART I

GETTING
Started

CHAPTER ONE

Deciding to Farm

Why do people farm?

Farming can be rewarding. There is nothing like the feeling that comes from walking through a still-foggy field in the morning to find newborn triplets from your favorite ewe, or seeing a foot-long ear on your latest strain of open-pollinated corn, or just looking across your land at sunset and saying, "This belongs to me!"

Farming can be frustrating. Why did that chain choose to break on your planter right in the middle of the farthest field at 3 P.M., forcing you to have to go to town to get another, only for it to be dark when you got back? Why did it rain for three straight weeks while you were trying to plant your vegetables, not raining a drop since? A man once told me, "There's no way in hell I'd have a job where my income depended on whether or not it rained!"

Farming can be tough. The number of farmers in the United States has shrunk from 6.5 million in the 1930s to only 2 million in the 1990s. Less than 12 percent of all farmers make a sustainable living wage from the farm, and we are losing almost one hundred farms per day. Agricultural experts (economists in particular) have been predicting the demise of small farms for the past 50 years. "Look before you leap" certainly seems like good advice if you are thinking about becoming a farmer.

◄ *Today's small farms range from quarter-acre specialty herb farms to many-acred traditional and alternative crop and livestock farms that sell value-added products directly to the consumer. Can you visualize your farm?*

3

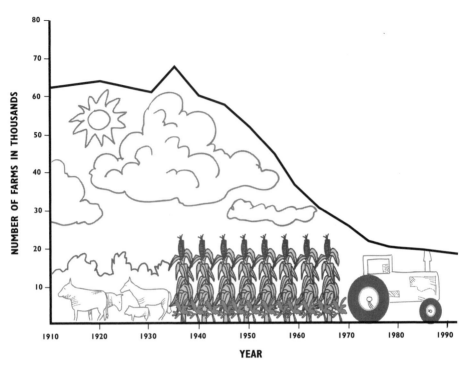

The total number of farms in the United States has declined from a high approaching 70,000 in the 1930s to 20,000 today.

So why do people farm?

The reasons that people begin to farm are as varied as the snowflakes that are just now beginning to fall on my newly budding trees. Some farmers are sons and daughters following in their parents' footsteps. Some have fond memories of spending summers on the farm of a grandfather or uncle. Some are influenced (as I was) by an elderly neighbor talking about his life on the farm.

The Rural Sociology Department at the University of Missouri did a study in the 1950s, asking why people chose to farm. Large and small farmers alike were questioned. Here were the first three reasons:

1. "I like to work outdoors."
2. "It's a good place to raise my children."
3. "I'll always have a place to live and food to eat." (Not necessarily true, if you have a mortgage.)

Fourteenth of the fifteen reasons listed was "It provides a good income," and fifteenth was "I don't know how to do anything else." Obviously, lifestyle was a big reason that people farmed, if income was so far down on the list.

Almost 40 years after the University of Missouri study, the DuPont Company hired Doanes Agricultural Service to do a study on why farmers farm and what farming practices they used, among other such questions.

In that study, conducted in 1996, the first three answers were still lifestyle choices, but "good income" moved up to seventh place. Income seems to be more important to people now than it was four decades ago, but lifestyle is still the principal reason someone farms.

Later in this chapter, I will explain the importance of farm income for small farmers, and why it should be your number one goal; the reason will probably surprise you. But even though you need a good income to be a farmer, that is probably not *why* you want to be a farmer.

This is also true of me. The reasons I farm all involve the independent lifestyle, self-reliance, being in harmony with nature, and the feelings of goodness that come from my land. Self-reliance and a love of the outdoors are an ingrained part of our American heritage, and farming is a way to celebrate that heritage. Henry Ford said it best: "The further removed we are from the land, the greater our insecurity."

What Is a Small Farm?

If you are planning to farm, you are almost certainly going to start with a small farm. You will be in good company — 79 percent of all the farms in the United States are small farms, according to the 1997 Census of Agriculture. There is also good news for small farmers — while the total number of farms has decreased, the number of small farms is actually increasing at a rate of about 2 percent per year, and this trend is predicted to continue for at least the next 10 to 20 years.

How much land do you have to have to be a small farm? The answer is up to you. I define a small farm as any farm that has 179 acres or fewer, or that has a gross income of $50,000 per year or less. This seems to be a reasonable compromise between acreage and income — but not everyone agrees. As you read this book, don't get hung up on numbers and definitions. There are hundreds of government bureaucrats and university professors making $100,000 a year arguing about what's a large farm and what's a small farm. To make the matter worse, many people confuse small farms, family farms, and sustainable agriculture. The U.S. Department of Agriculture has no official definition of farm sizes. Meanwhile, according to the National Small Farm Commission, a small farm is any farm grossing $250,000 per year or less. This is 94 percent of all the farms in the United States, and includes what I would call midsize farms.

There even seems to be some argument as to what constitutes a farmer. The first definition of a farmer in Webster's *New*

TERMS

The confusion between small farms, family farms, and sustainable agriculture is somewhat understandable, because all three share similar characteristics.

❖ A *small farm* is any farm that comprises 179 acres or less, or that grosses $50,000 or less per year. Small farms are usually family farms but may or may not be sustainable.

❖ A *family farm* is any size farm — small, medium, or large — in which family members supply the majority of needed farm labor. A family farm is not necessarily sustainable.

❖ *Sustainable agriculture* is an economically viable, environmentally sound, and socially acceptable system of agriculture that may be used on any size farm.

World Dictionary is "a person who earns his living by farming; especially, one who manages or operates a farm." While Webster was content to define a farm and a farmer using only words, the rest of the world seems to prefer definitions involving numbers. The USDA has changed the definition of a farm nine times. In the 1950s and 1960s, a farm was any place smaller than 10 acres but selling $250 or more of agricultural products per year, or any place of 10 acres or more from which $50 of agricultural products were sold per year.

Currently, the USDA defines you as a farmer if you sell $1,000 worth of agricultural products per year. In 1996, the U.S. Bureau of the Census proposed changing that figure to $10,000 (presumably to save itself some counting). There was an outcry from universities, because the sizes of their federally funded programs are determined by the number of farmers living in their state; in the end, the $1,000 figure has prevailed.

My own definition of a farmer is quite simple: If you think you're a farmer, you are a farmer. Any size acreage, from a small garden to 3,000 acres, where a person or family tries to make a living (or part of a living) from the land is a farm. A small farm is any farm that is small, regardless of cash inflow or outgo. The "Biggest Little Farm in America" grosses $238,000 from ½ acre of gourmet vegetables. I know several owners of 80-acre farms who gross $50,000 to $60,000 per year, with a final (net) profit of $25,000 to $30,000. On research plots on my 80-acre farm, with a combination of vegetables and livestock, we have grossed $3.00 per square foot, and netted $1.32 per square foot. If we expanded these practices to a full acre (43,560 square feet), this would yield a gross profit of $130,680 and a net profit of $57,500.

The definitions of farms and farmers probably always will be contested by a variety of university, corporate, government, and individual entities. If we are to progress as farmers, these definitions are not necessarily important. We must simply know who we are, what we are about, and where we are going. Agricultural problems are solved by solutions, not definitions.

How the IRS
Defines a Farmer

The U.S. Internal Revenue Service (IRS) has its own definition of a farmer. Because farmers, unlike hobbyists, are allowed tax reductions, the IRS defines them rather strictly. The short explanation (available in "Farmer's Tax Guide 225") is that you may file as a farmer by March 1 if at least two-thirds of your gross income is from farming. Further, your farming activities are presumed not to be a hobby if profits result in any three of five consecutive tax years, ending with the tax year in question — with exceptions for raising horses. Obviously, the IRS is not interested in unsuccessful farmers.

A more detailed explanation is available in publication N-1000, "Farmers for Tax Purposes" (italics are mine):

> Who is a farmer: *All individuals, partnerships, or corporations that cultivate, operate, or manage farms for gain or profit, either as owners or tenants, are regarded as farmers.* To be classified as a farmer, a taxpayer must meet the two-fold test of participation to a significant degree in the growing process and assumption of a substantial risk of loss from that process. Thus, a taxpayer who gets rent based on farm production is a farmer. But if the rent is fixed, he isn't a farmer unless he materially participates in the farm's operation or management. A taxpayer is engaged in the business of farming if he belongs to a partnership so engaged. . . . *Farmers include persons engaged in oyster farming, the raising of bees, breeding and raising chinchillas, mink, foxes, and other furbearing animals.* Feedlot operators have been held to be farmers with respect to stock they own, as well as with respect to stock handled for customers. *But a taxpayer engaged in forestry or the raising of timber is not engaged in farming. Neither is a taxpayer who sells Christmas trees grown, without planting or cultivation, on land he owns.*

Well, that really clears things up! This is why I call myself an agripreneur, instead of a farmer.

Becoming a Farmer

Farming, in some respects, is easy; making a living from a farm is hard. In addition to being a farmer, I want to help you become an agripreneur.

Agripreneur is a term I coined in 1987 to describe the readership of my magazine, *Small Farm Today*. An entrepreneur, according to Webster's dictionary, is someone who runs a business at his/her own financial risk — a middleman. An agripreneur is someone who runs an agricultural business — farming in particular — at his or her own risk.

I will discuss this in more detail in later chapters, but for now, ask yourself some simple questions:

- ❖ "How much money do I need in order to live comfortably and support my family? $15,000? $20,000? $50,000?"
- ❖ "How long will it take to achieve this level of income, and will I be happy when I achieve it?"
- ❖ "Do I want to farm part time? Full time? Start part time and grow from there?"
- ❖ "How much money do I already have in order to start farming, and how much do I wish to borrow?"
- ❖ "What skills and resources do I have? Am I good with livestock, or can I learn to be? Am I good with machinery, or maybe carpentry — to build hog houses or poultry houses? What skills do I have that I really enjoy? What skills am I weak in or do I really dislike?"

SELF-EVALUATION

Becoming an agripreneur involves changing the way you think about farming. When you decide to start a farm, you need to ask how farming will satisfy both your monetary needs and your personal goals, now and in the future. This is a question that you will want to reevaluate frequently as you continue in farming.

These are not questions you will know all the answers to now, but you need to keep them in mind as we proceed.

Do not be critical of your skills at this point. There are many people entering farming today with no farming background. These people have good business and computer skills — which can be a farming asset — but don't have any of the basic day-to-day knowledge to run a farm, from milking a cow to planting corn. If you are one of these people, don't worry; you can learn.

Full-Time Farming: Pros and Cons

If you want to start out as a full-time farmer and have no experience, you have chosen a difficult road; you will have to do all your learning while you are striving to make enough money to satisfy your personal goals. It is even harder if you took out a mortgage and a big loan; on a full-time farm, your business, your home, and your lifestyle are all tied together in one big package.

On the other hand, in full-time farming you are your own boss and set your own hours. You are the CEO of planning, developing, and determining your destiny. It is highly rewarding to lay out a plan and guide it to completion, correcting your failures and overcoming obstacles. No amount of money can give you the feeling you get in the springtime from watching the birth of a scruffy bright-eyed calf or a litter of tiny piglets, as your farm renews and repopulates itself in its never-ending cycle of life.

Part-Time Farming: Pros and Cons

Part-time farming allows you to have your cake and eat it too. Combining farming with a town job or seasonal off-farm labor can be difficult to manage, juggling the demands of both workplaces, but you'll have more financial safety. It is certainly a safer transition for new farmers; their farming mistakes will not have the capability to cost them their entire income. An off-farm job provides the financial security and the borrowing power of a steady paycheck, and reduces the risks involved in a weather-dependent business.

You can experience the rewards of farming and learn as you go without the stress of a "root-hog-or-die" existence.

Gaining Knowledge

I was lucky enough to have some knowledge when I started. I grew up in a rural area, although my father did not farm. I worked on several farms nearby and learned skills as varied as driving a team of mules to harvest a potato crop, raking and baling hay, and milking a ten-cow herd by hand for grade C milk. If you are new to farming and you know any farmers, talk to them and do chores with them whenever you get the chance.

People who grew up on farms do not always discuss their finances in depth, nor do they necessarily know why they do things the way they do. They may simply be carrying out the instructions they learned about *how* to do it, without understanding the principles of *why* they do it. To be a successful agripreneur, you must understand your operations completely, from the how to the why.

The Farming Life

PRINCIPLE: *Farming goals must be family goals. To move to a farm, the entire family needs to be prepared for what life will be like.*

Although life on the farm is much easier than it was in the past — due to the advent of electricity, county water, and the automobile — a farm is still a rural environment, and when it gets dark, there are no streetlights. Dust and mud are "up close and personal" every day and every season. Dirt and gravel roads are harder to manage than concrete, and no city crew comes out to clean snow off your street. You will probably be far away from medical help, a fire department, and the police. Your children will have fewer playmates, and visiting the neighbors may require hopping into a car.

On a farm, you'll have to deal with nature. Nature is cyclical, slow, and unpredictable. After a beef cow is bred, she is pregnant for 9 months, and then you must wait another 6 months before you can sell her feeder calf to get back any income. Many field crops take 6 months to a year from planting to harvest to income. The weather can act up at any time, drowning your crops or freezing your livestock.

If you are new to farming, all of this will require some adjustment. Fortunately, there are genuine advantages to a rural life. People in a rural environment learn to become more independent — to rely on their own inner strength. Family members who work together and share will learn to depend on each other much more than they would in an urban environment, where everyone is going in different directions.

The most important thing that agriculture has furnished this country with is not food or fiber, but, rather, a set of children with a work ethic and a good set of values. Doing farm chores that must be done — it is not acceptable to say, "Oh, I'll feed the goats next Saturday" — gives children a basis for a work ethic that will continue throughout their lives. These values carry over into the rural community, creating a quality of life that we all want — and need.

Children who grow up on a farm learn responsibility and reliability.

One of the biggest differences between a town business and a farm is that at the end of the day, you leave a town business and go home. With a farm, your business, your home, your family, and your lifestyle are all linked. Dinner-table discussions on a farm usually include a lot more "business talk" than do dinner-table discussions in town. And when your home and farm are at the same location and your mortgage entangles both, if your farm is not profitable, you can lose both your home and your business in one fell swoop.

On the bright side, watching the sun set after a hard day's work is much more satisfying from your porch than it is from rush-hour traffic. And to start the day's work, all you have to do is step outside instead of stuffing your briefcase and jumping into your car.

None of the obstacles is hard to overcome. The important thing is for the entire family to be prepared for a lifestyle change. Husband, wife, and children must know why they are making this choice and what part they will contribute to the whole.

Agripreneurship

Sustainable farming is the current catchphrase in agricultural circles. Sustainable farming simply means farming that sustains itself — a continuous cycle that does not wear out the land or the farmer, replenishes the livestock and crops, and enables the family to continue farming. There are many elements to sustainability, such as diversity and "low input" — but the most important principle of sustainability is to be profitable. Memorize this sentence: "To be sustainable, it must be profitable."

This is important for the survival of your farm and is part of my definition of agripreneurship. Apripreneurs must have a positive attitude and practice a sustainable type of farming that satisfies both personal and family goals. Because agripreneurship involves sustainability, and sustainability requires profitability, we will make this the guiding principle of agripreneurship.

A Profitable Farm

PRINCIPLE: *Your farm must be profitable.*

Many people believe the goal of being profitable is to get wads of money so they can spend it on what they want. These people pursue high-paying jobs and put in endless hours of work to get rich, so they can be happy.

The reason to make your farm profitable is simple: As long as you take in more money than you spend, your farm can improve in fertility and has a potential for sustainability. A farm can generate large numbers of bills — your goal is to be able to pay them.

If your farm has to be subsidized with off-farm income, it is not sustainable in the long run. Eventually you'll deplete your savings, or you will retire from your job. Even part-time farms should be sustainable. Make your farm pay for itself.

In today's small-farm market, your only real financial security lies in your ability to sell yourself and your products. You will be responsible for your own success or failure. In conventional agriculture, a farmer simply raises the crop and then hauls it to the middleman to sell. The middleman passes on the crop to the retailer and eventually to the consumer. To be an agripreneur, you must be your own middleman, selling your product directly to the consumer.

The Big Picture

PRINCIPLE: *You must look at the whole picture.*

Agripreneurship involves examining your operation as a whole. Its parts should support each other, and you need to be aware of how efficiently each part is succeeding — or failing.

In a factory job, workers simply screw part A into part B and never see the whole. In most office jobs, each worker gets a piece of a project, or a single account out of several. An agripreneur must be aware of each piece, each account, and how they are linked. Your farm is the entire factory, the whole company, and you need to know how all of it works.

To see the whole picture, you must look beyond your farm, too. Is your product something you can sell in your area? Are there "new" products that would be appreciated in the local community? How does what you raise affect neighboring farms? How does what they raise affect you? What will be the overall effects in the local region?

Planning

Agripreneurship is the "thinking person's agriculture." You must plan carefully, for both your farm and your family. Your plan should include time for rest and recreation, because this renews your spirit and gives you time to think about your farm and learn what others do. Farming is physical by nature, but it is not how hard you work, but how smart you work. The difference between a conventional farmer and an agripreneur is that the agripreneur does more thinking and less doing. If you wear yourself out physically and mentally, you will accomplish nothing. Learn to enjoy the trip, as well as the destination.

In agripreneurship, your reasoning process, backed by reading and research, enables you to make decisions that affect or contribute to all facets of your farm and your personal and family goals. A thinking person evaluates all of his or her options and tries to keep an open mind. My favorite maxim is, "Just because it's new doesn't make it better, and just because it's old doesn't make it obsolete." An agripreneur will take the most useful of the new with the best of the old and apply it to his or her own farm in an individual style. To discover what will work and what won't requires research, experimentation, and thought.

❖❖❖

FOOD FOR THOUGHT

There are other principles to both sustainability and agripreneurship, but this will get you started. Remember that agripreneurship — and farming — is a never-ending process of discovery. Your farm will evolve into a unique entity and meet your personal goals, family goals, and farm goals. Once you grasp and apply these principles, you can become independent, self-sufficient, successful, and happy. Just think of this as your agricultural self-help book.

❖❖❖

Starting a Farm

There are many different types of farms, from large monoculture farms that produce a single product to general diversified farms that raise a little bit of everything (see chart for examples). Farms can range in size from thousands of acres down to ½ acre or even smaller. A shiitake mushroom farmer in southern Missouri got her start with eight mushroom logs. Now she farms thousands of logs.

The important thing to consider when choosing a farm type is to find something you like to raise that is compatible with your climate and that can be marketed for a profit. We will discuss this further after you have done some farm planning.

Starting a Farm Plan

Whether you are buying a new farm or jump-starting an old farm, you must have a plan. To begin your farm planning, first assess your available resources, both personal and property.

If you are new to farming, there is a definite advantage to deciding what you will farm (vegetables, grains, dairy cattle, ostrich,

≺ *Farming furnishes children with a work ethic, a good set of values, and a true sense of family*

Some Types of Farms

Farm Type	Product/Characteristics
Traditional crop	Field crops such as corn, soybeans, wheat, and milo
Crop and livestock	May use rough ground for pasturing livestock and the rest of the property for traditional field crops
Specialty crop	Herbs, cut or dried flowers (outdoors or in greenhouses), industrial field crops such as guayule (a rubber substitute) and kenaf (used as fiber by the paper industry), one or several varieties of vegetables, fruits (from kiwis to apple orchards), and berries; fruits and vegetables can be marketed as either fresh produce or value-added crops (canned tomatoes or dried blueberries or jam, for example)
Unusual crop	
• Butterflies	Live butterflies, for use at weddings and parties
• Bees and insects	Rent insects out for pollinating crops
Aquaculture	
• Worms	Worms for fish bait
• Goldfish and tropical fish	Fish for the aquarium trade
• Fee fishing	Consumers pay a daily fee or a per-pound fee for fish they catch on the farm
• Food fish	Catfish, trout, and other fish are raised for sale to stores, restaurants, and consumers
Exotic animals	
• Elk, bison, deer, mountain sheep, donkeys, ratites (e.g., ostriches, emus, and rheas)	Raised for meat, hides, and specialty by-products such as emu oil or velvet from elk antlers (used medicinally)

for example) and how you will market it (roadside stand, farmers' market, sale barn, value added) before you purchase the land. On the other hand, it is difficult to find the perfect farm that has everything you want at a price you can afford. So whether it is a new farm or an old one, you need to evaluate the property. By *property*, I mean not just the land you own, but also what you will place on it.

Evaluating Your Resources

Make a big resource list (see box on page 20) — try to include on it everything you can think of. A list of what you know and what you need to know is the first step to a successful farm. If you know what you have available, it will make it easier to work out what is required for whatever enterprise you choose.

Farm planning is like a road map: You must know where you're going, or it will be awfully hard to get there. In other words, you must know how many dollars you need to live, both now and in the future, and how many potential dollars a farm operation can expect. It is obviously not possible to come up with exact numbers, but you must have some realistic figures in mind. You began evaluating your capital and skills in chapter 1, in the Self-Evaluation section, but let's consider them more closely now.

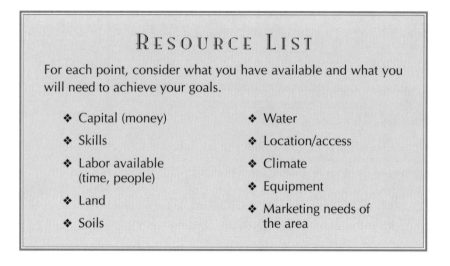

RESOURCE LIST

For each point, consider what you have available and what you will need to achieve your goals.

- Capital (money)
- Skills
- Labor available (time, people)
- Land
- Soils

- Water
- Location/access
- Climate
- Equipment
- Marketing needs of the area

Capital

In chapter 1, you evaluated what income level would be comfortable for you and your family. You also began to consider how much money you would need to start farming, or to change your operation. Now let's think about the amount of money you currently have. Ask yourself these questions:

- Do I have any savings?
- How much am I willing to risk on this new farming project?
- How much can I borrow — and how much am I willing to borrow?
- Can I rent land with or without the option to buy later to maximize my limited capital? Should I rent or should I buy?
- Do neighbors or nearby equipment dealers have machinery I can rent, rather than buy?
- Can I buy a cheap "fixer-upper" and use my knowledge and time to multiply some sweat into cash?

These options all arise from the direct monies you have available. Consider other means to gain what you need as well, such as barter. Barter can be expressed in many ways. You could farm a piece of land on shares — two-thirds to the owner, one-third to you. You can trade crops and livestock back and forth between nearby farms, or offer them in trade for a fence, or hay baling, or some plywood for a sheep shelter. Your skills — or just some honest sweat helping another farmer — are also tradable. Only your imagination limits the capital you can get.

Skills

In chapter 1, you began a list of your personal resources, including the skills you have available. By listing the skills you have, then evaluating those you will need for your farming operation, you can determine what you will need to learn, or what you need someone else to do in your place.

The only way to acquire all the basic skills you need to be a farmer is to farm. Although you can (and should) read a book to learn the theory of how to plow a field, in reality there are far too many variables to explain. If you are completely new to farming, it may be wise to work for another farmer for a while, helping him in return for being taught some basic how-tos. Workshops, seminars, classes, and conferences are other places to learn at least the rudiments of many skills. If you cannot find anyone to learn from, read about it, then just jump in and do it. If you're patient, eventually you'll get it right — as they say, "Experience is what we learn from our mistakes."

Keep in mind that agriculture can be dangerous, especially when lacking basic livestock- and machinery-handling skills. Don't get injured before you really get going just because of pride. Make sure you have a good idea of what you are doing, and make sure you have the proper tools to do it.

Many old farmers would be glad to teach you their skills. Talk to them about volunteering in exchange for instruction, at sale barns, at feed and supply stores, or at their farms.

Skills a Farmer Needs

Following are lists of skills that are needed for some basic types of farming. Of course, you might research what is needed for the operation you are planning. For example, elk farming, although similar to cattle farming, has very different requirements for each skill, and knowledge of collecting velvet (a product of the antlers) will also be necessary.

Some of these may not be as simple as they appear. Moving cattle, for example, requires knowledge of the best places to stand to appear in the animals' fields of vision without alarming them, and what types of motion and movement work best.

Field-crop skills
Soils
Plowing, disking, harrowing
Planting, cultivation, harvesting
Skills for each piece of equipment used
Weather/season knowledge
Crop knowledge
Pest control, weed control, fertilization
Seed saving/preparation
Irrigation
Marketing

Vegetable-crop skills
Soils
Crop knowledge
Field preparation
Planting, seeds, transplants
Vertical growing, trellis (e.g., peas), cages (e.g., tomatoes)
Raised-bed growing
Flat-field growing
Mulching, cultivation
Harvesting
• Hand
• Mechanical

Vegetable-crop skills (cont'd)
Weather/season knowledge
Use of season extenders
Pest control, weed control, fertilization
Seed saving/preparation
Irrigation
Marketing

Vine- and bramble-crop skills
Soils
Planting
Propagation
Identifying new wood, pruning, trellising
Cultivation
Harvesting
• Hand
• Mechanical
Adding value
Winter preparation
Weather/season knowledge
Crop knowledge
Pest control, weed control, fertilization
Irrigation
Marketing

Cattle skills

Applying ID tags
Castration
Breed knowledge, breeding
Dehorning (removal of horns),
    hoof trimming
Medication application,
    shots with needle,
    worming
Delivering calves
Milking cows
Handling/moving cattle
Restraining large animals
Behavioral knowledge/control for:
  • Aggression
  • Birthing
  • Sickness
Shelter needs
Feeding/water
Manure handling
Fencing
Marketing

Hog skills

Breed knowledge, breeding
Flushing
Applying ID tags
Castration, ringing
Restraint for health care
Birthing of litters
Handling/moving hogs
Restraining large animals
Behavioral knowledge/control for:
  • Aggression (boars and sows)
  • Birthing
  • Sickness
Shelter needs
Feeding/water

Hog skills (cont'd)

Manure handling
Fencing
Marketing

Sheep skills

Breed knowledge, breeding
Flushing
Applying ID tags
Tagging (removal of wool around
    udder), docking (removal of
    tails)
Medication application
  • Shots with needle
  • Worming
Delivering multiple lambs
Shearing
Handling/moving sheep
Behavioral knowledge/control for:
  • Aggression (rams)
  • Birthing
  • Sickness
Shelter needs
Feeding/water
Manure handling
Fencing
Marketing

Useful hand-tool skills

Woodworking
Concrete working
Fencing
Electrical
Chain-saw operation
Rototiller usage
Hand mowers

Labor

You must assess your available labor, in terms both of yourself and of your family. Will it be just you working, or will you have some combination of family members? If you have an idea of what combination of crops and livestock you plan to raise, figure out whether you have enough labor to take care of these enterprises. Remember, as we discussed in chapter 1, your financial goals must be your whole family's goals in order to be successful.

Farm labor is generally measured by the number of 10-hour work days required under average conditions to take care of a certain number of acres or so many head of livestock (see labor charts, pages 26 and 27). Six 10-hour days per week for 50 weeks allows about 3,000 hours of labor available per year for one full-time farmer. In general, increasing the number of labor hours means increasing income, but in the case of a thinking agripreneur, this is not always so.

FAMILY LABOR

Family labor makes use of each family member's skills in a way that yields the greatest benefit to the farm. Each of us is better at some things than others, and tasks should be divided accordingly. Try to avoid strict rules and stereotypes: If a woman is better at lambing ewes, and a man is better at keeping books, do it that way. Even small children — with patience, instruction, and careful supervision on your part — can be helpful. Small children can handle baby chicks, fill feeders or water pans (a little at a time), pull weeds, or pick small fruits once you show them how. Give older children more responsibility, and listen to their ideas. If you want children to develop a strong work ethic while learning useful skills, they must first enjoy their work. You must teach them how. (See chapter 10 for more on family labor.)

Using the amount of hours listed in the labor charts, it seems that one person could therefore take care of 428 beef cattle (7 hours per cow), 750 sheep (4 hours per ewe), or 7,500 laying chickens (40 hours per 100 hens) — although this would not allow time for building or equipment maintenance or marketing. A better time allowance (though still far from ideal) on a diversified family farm might be 20 acres of corn (152 hours), 20 acres of grass/legume hay (218 hours), 420 acres of pasture (29.4 hours), 100 beef cattle (700 hours per year), 100 sheep (400 hours per year), and 2,000 laying hens (800 hours per year). This would total 2,300 hours per year, leaving 700 hours to use on maintenance, making hay, marketing, or a part-time job.

When allotting time for labor, remember to allow for the proper distribution of labor throughout the year, so that you are not crowded for time during critical months. Field corn, for instance, requires 7.6 hours per acre per year — but 1.1 hours of that are in December to March, 3 hours (planting) are in April and May, 1 hour (cultivation) is in June and July, 0.8 hour is in August and September, and 1.7 hours (harvest) are in October and November. The heaviest months are April through May and October through November. In contrast, in a two-litter-per-sow Butcher Hog system, with litters farrowed on March 1 and September 1, 16 hours are required in December to March, 8 hours in April to July, 4 hours in August, 8 hours in September, and 4 hours in October and November. Note the increased labor hours at farrowing time.

Because your goals should be your family's goals, it is important to set aside time for family in your farming life. Don't let your work schedule get so full that you neglect other obligations, such as time for church, vacations, or children's sports events. Remember, too, that labor is not accomplished solely on the farm. Allow time for education: Attend seminars and conferences. Nonfarm activities give you personal contacts with different viewpoints and may inspire ideas about how to improve the management of your own business.

LABOR CHARTS

These charts give an estimate of time required for all direct labor involved in the normal care of crops and livestock: planting and harvesting, feeding, milking, cleaning stalls and pens, castration, docking, assisting in birth, and normal health care such as shots and vaccinations. They do not include establishment of a product (start-up times), machinery and building maintenance, cleaning bins and other general farm chores, adding value, or marketing of the product. The labor hours are based on the types of equipment a small farmer could be expected to own (a three-plow tractor, for example). These times are approximate and should be used only as guidelines in comparison to your actual records.

TOTAL HOURS OF LABOR PER YEAR TO PRODUCE THE CROP AT AVERAGE EFFICIENCY FROM FIELD PREPARATION TO HARVEST

Crop	Hours/Acre	Crop	Hours/Acre
GRAINS, GRASSES, AND LEGUMES		FRUITS AND VEGETABLES	
		Beans (snap)	109.2
Alfalfa hay	14.8	Broccoli	136
Barley	3.8	Cantaloupe	108.9
Clover hay	6	Corn (sweet)	71.7
Corn (field)	7.6	Cucumber	153.2
Corn (silage)	11.4	Eggplant	207.8
Cotton	20	Okra	301.8
Grass hay	5.4	Pepper (bell)	185
Grass/legume hay	10.9	Potato	105.6
Oats	3.3	Raspberry	100.7
Pasture (crop rotation)	2.4	Spinach	103.4
Pasture (permanent)	0.07	Squash	140.7
Sorghum	7.6	Strawberry	140.3
Soybeans	5.2	Tomato	306.4
Soybeans (second crop)	4.2	Watermelon	76.5
Wheat	3.8		

Data from *Farm Business Planning Guide for Organization* #6500, University of Missouri Cooperative Extension, Columbia, MO, 1965; *Selected Fruit and Vegetable Planning Budgets*, EC 959, by Charles D. DeCourley and Kevin C. Moore, Department of Agricultural Economics, University of Missouri—Columbia, 1987; and *Enterprise Budgets: Northeast Oklahoma 1985*, prepared by Bill Burton, Oklahoma State University Cooperative Extension, Claremore, OK, 1985.

TOTAL HOURS OF LABOR PER YEAR TO
RAISE ANIMALS AT AVERAGE EFFICIENCY

Animal Unit	Hours/Animal Unit
Butcher hogs: 1 sow (2 litters/year)	40
Feeder pigs: 1 sow (2 litters/year)	22
Feeder pig finishing: 100 pigs	80
Sheep: 1 ewe	4.0
Honeybees: 1 hive	6.2
Dairy goats: 1 doe	1.2
Chickens: 100 laying hens	40
Chickens: 500 broilers	9
Turkeys: 100 birds	10
Dairy cattle: 1 cow	
Fluid milk market	80
Manufacturing milk market	85
Replacement heifer to 2 years old	20
Beef cattle: 1 cow and calf	
• Stocker calf	7
• Steer calf	
Wintered only	3.5
Wintered and grazed	4
Finished immediately	5
Wintered and finished	6
Wintered, grazed, and finished	8
• Heifer calf	
Finished immediately	4
Wintered and finished	6
• Yearling steer	
Finished immediately	5
Wintered and finished	6
Plain wintered and short fed	5

Data from *Farm Business Planning Guide for Organization* #6500, University of Missouri Cooperative Extension, Columbia, MO, 1965, and *Enterprise Budgets: Northeast Oklahoma 1985*, prepared by Bill Burton, Oklahoma State University Cooperative Extension, Claremore, OK, 1985.

> ## Terms
>
> **Finished immediately** Fed only grain after weaning
> **Heifer calf** Female who has not yet given birth
> **Steer calf** Castrated bull calf
> **Stocker calf** Calf sold at weaning (6 to 7 months of age)
> **Wintered only** Weaned from mother and fed hay or grain over winter
> **Wintered and finished** Fed hay and grain until spring, then given all the grain the animal desires to finish
> **Wintered and grazed** Fed hay or grain over winter, then grazed on spring grass until sold (usually in autumn)
> **Wintered, grazed, and finished** Fed hay and grain after weaning, grazed from spring to autumn, then finished on more grain until marketed
> **Yearling steer** Steer about 1 year old; a "short" yearling is just under a year, a "long" yearling is just over a year.

Land

Questions to begin assessment of the property include:

- ❖ What size is my farm?
- ❖ How much of the land is producing timber or brush?
- ❖ How much is suitable only for pasture and how much is suitable for grain crops?

This can be further divided into small-grain-crop (wheat, oats) and large-grain-crop (corn, soybeans) suitability, based on soil types. Large grains usually require more fertile soils.

When buying property, remember that real estate is a piece of land with a set of rights that may or may not be included. Make sure you have water rights and mineral rights, and are aware of easements and any other intrusions on the property. Check at the Abstract Office or Title Company in your town to see what rights have been sold.

Condition of land, soil type, and climate are the main factors in determining what crops will grow on an area of land.

Although it is possible, in fact, to grow almost any crop anywhere, the cost to grow some crops in some areas may not be acceptable. Grain and vegetable crops do best in deep, fertile soils and level fields. Fruit crops do best where there are no extremes of temperature. They need well-drained soil, but do not require a level topography. The condition of a land is also known as its capability.

Classifying Land

Land is generally classified into eight Capability Classes. Each class allows for certain types of usage and treatment. Classes are based on slope of land, available water capacity, and soil drainage. Classes I to IV are suited for cultivation. Classes V to VIII may be pasture, woodland, or wildlife areas. These are general conditions, and the land may be capable of alternative uses with proper treatment. For example, a hillside may be used for crops if water is prevented from running straight down the slope and eroding the soil. Terracing (a series of steps or ridges made from soil running at right angles to the direction of slope), contour strip-cropping (alternating strips of six to twelve rows of two different crops, forming bands following the contours of the land), and crop rotation (alternating crops by year or season to provide better soil and more ground cover) can make a hillside viable for crops. Your Soil Conservation Department can provide you with maps of land capability. Check with your university extension office or your state Department of Agriculture.

Class I land, the best possible, generally has a slope of 0 to 2 percent, with deep, fertile soil that does not erode easily. It holds water well, has a good supply of plant nutrients, and is free of rocks. It is good for any use, from cultivated crops to pasture, from woodland to wildlife.

Class II land, usually with a slope of 2 to 5 percent, is almost as good, but some conservation practices may be necessary to keep it productive, due to erosion, wetness, or lack of water retention.

Class III land, with a 5 to 10 percent slope, may require intensive erosion control if regularly cultivated. It may be excessively wet in spring or poorly aerated.

Class IV land is still suitable for cultivation, but only for limited periods, and may be better used as pasture. It usually has a slope of 10 to 15 percent.

The other four classes are not generally suitable for cultivation. **Class V land,** with a slope of 15 to 20 percent, is limited to plants that can cope with extremely wet, poorly aerated, rocky soils. **Class VI land** is suitable for pastureland or woodland, with limited conservation practices. It may have shallow soil, steep slopes, high erosion, or lots of rocks. **Class VII land** is good for wildlife, but may have severe limits for pasture or woods. **Class VIII land** won't grow much of anything.

Soils

> **PRINCIPLE:** *Natural fertility and slope of land are critical on small acreages.*

There are more than 18,000 different soils in the United States. They vary from shallow to deep, from clay to sand to loam, from well drained to wet. Each type of soil is suitable for particular kinds of crops or land usage. For example, the Clinton-Boone-Lindley areas in Wisconsin, Minnesota, Iowa, Illinois, and Missouri vary from gentle to sharply rolling hills. Most of this region is uncleared woodland or pastureland. Cultivated areas are small, and crops depend on slope. This is not a good soil type on which to plant 400 acres of corn, but it might make excellent orchard land.

There will be more on soils in chapter 4, but you need to know the natural fertility of your farm soils and roughly what condition they are in currently. For any farm of less than 10 acres, the natural fertility of the farm is very important, because the physical size of the land will restrict the choice of enterprises. For instance, a cow-calf operation, which takes 3 acres to run a cow-calf pair, will not allow for much volume of income. The poorer the soils, the more area it would take; in some areas of the West, it takes 10 acres to run a cow-calf unit. Ten acres of vegetables or U-pick berries, however, would produce a large volume of income — but these require adequate water. Each enterprise must be evaluated in the light of the current resources and conditions.

A good place to start your soil research is with soil maps of your area. Contact your university extension office or state Department of Agriculture to find the nearest map source, or consult the appendix for the national offices.

Water

Because just one cow drinks 50 to 90 gallons of water a day, you can see how important water is to the farm. If you're going to raise crops and livestock, you need lots of water. We will cover this in more detail in chapter 5, but when you are thinking about buying your farm, find out what water resources you can expect to have. And be sure to ask these questions:

One cow drinks 50 to 90 gallons of water per day.

- ❖ Does the property have a creek or river running through it?
- ❖ Is this water source unpolluted?
- ❖ Where does the water come from?
- ❖ Does it start on my farm?
- ❖ Does the property have any ponds (or "tanks," as they are called in Texas)?
- ❖ Can cattle, sheep, or horses water from these ponds, or are they too shallow, with too much mud and vegetation around the edge?

Water plays a definite role in what crops can be grown. Small-grain crops like wheat are generally grown in areas of 15 to 25 inches of rainfall per year. Large-seed grains, like corn, are grown in areas having at least 20 inches of rain from April to September and a 150-day growing season with hot days and nights.

WATER REQUIREMENTS OF CROPS

Crop	Lb Water per Lb Dry Matter (average)
Alfalfa	831
Barley	534
Clover, red	789
Clover, sweet	770
Corn	368
Cotton	646
Milo	328
Oats	597
Potatoes	636
Rye	685
Sorghum	322
Wheat	513

Data was gathered in Colorado. This table is for general comparisons only; measurements vary by region and climatological conditions. One Eastern U.S. study, for example, found that corn required only 271 pounds of water per pound of dry matter.

Look into how water is supplied by the county, too. Many rural areas have what is known as *rural water,* which means a large number of rural residents are served by one drilled well with treated water. Check your water rights (the rights to access water flowing across and underneath your land) at the county court-house. Out West in arid regions, if you do not secure your water rights with the property, you may not even have drinking water. Don't just assume you can use the water on your land because it was that way where you grew up.

If the property has a well, find out all you can about the amount of water the previous owners used — and how they used it. Will it fit your needs? Check with the State Geological Survey for information on the well's water flow and quantity (gallons per minute). Ask about reliability during drought or normal dry weather. Water quality is also important. Have the water tested for mineral content and pollution. Taste the water. Is it high in iron or sulfur, or otherwise unpleasant to taste or smell?

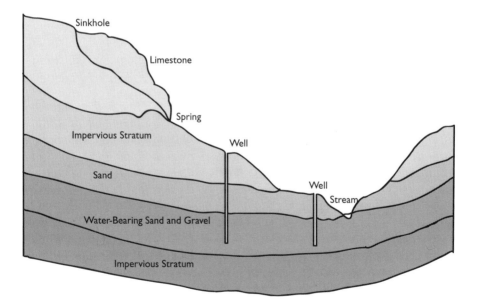

Wells must reach a permeable stratum where water flows freely, allowing easy access from the surface. The stratum can vary widely in depth in different regions.

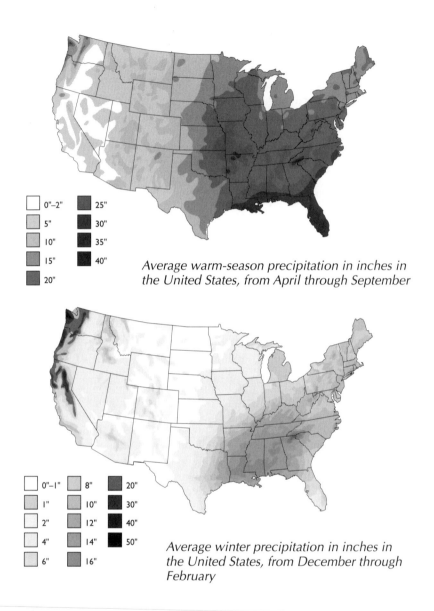

Average warm-season precipitation in inches in the United States, from April through September

0"–2"		25"	
5"		30"	
10"		35"	
15"		40"	
20"			

0"–1"		8"		20"	
1"		10"		30"	
2"		12"		40"	
4"		14"		50"	
6"		16"			

Average winter precipitation in inches in the United States, from December through February

Location/Access

Different areas of the United States are noted for different crops and livestock due to their topography (mountains, rivers, plains, for example), soil types, and climate. Kansas is known for wheat production, and Texas and Missouri for cattle. (Texas and Missouri have the largest number of small farms in the United States.)

Choosing the part of the country where you want to live will influence the type of farming you can do. You need to match the type of farming you are interested in with the soils and climate of the area. Look at the growing conditions of each crop and animal you are interested in, and determine if they are suitable for an area.

Other factors, such as livestock availability, must also be planned for; if you want to raise hogs in the western mountains, for instance, breeding stock will be many, many miles away. Distance from markets must also be considered. To direct-market your produce, you should be within 40 miles of a fairly large population. This is also important if you plan to sell some of your products through restaurants and grocery stores, for example.

Are you considering a U-pick operation or some form of agritourism? In order to sell produce on-farm or through a roadside stand, you need to be near some well-traveled roads, perhaps on the way to a tourist site, such as a boating lake. Ask yourself how many roads lead to your farm. Are they blacktop? Are you on the blacktop or a major highway or are you back in the "boonies"?

Easy access is important for hauling your crops and livestock to town, even if you don't have a U-pick operation. How close are you to a large-population town where you can purchase all the products and input items you need on the farm? Access to the farm is important both for visitors and for your family. (See chapter 7 for more information.)

Climate

Climate relates to your farm location. The USDA has issued a map of climate zones for growing plants. Zone designations are based on the minimum weather temperatures in that region, which determine the general hardiness of plants able to grow there. There are eleven zones for the United States.

In general, the lower the zone number, the hardier your plants must be to do well. In Zone 3, the first zone in the forty-eight contiguous states, plants must be able to endure –40 degrees Fahrenheit

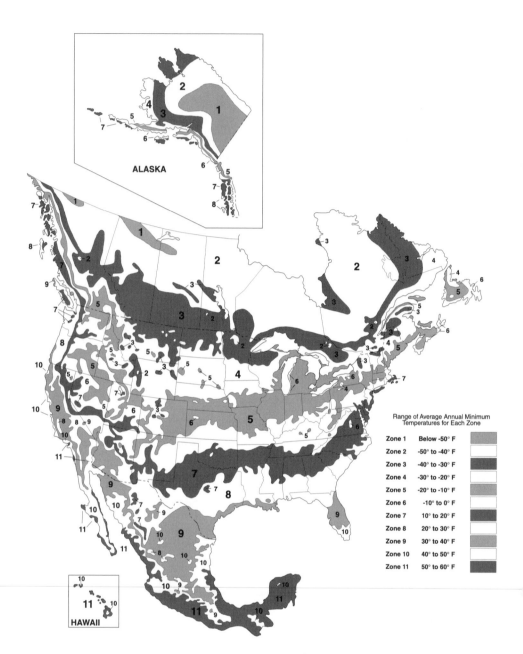

ALASKA

HAWAII

<table>
<tr><td colspan="3">Range of Average Annual Minimum
Temperatures for Each Zone</td></tr>
<tr><td>Zone 1</td><td>Below -50° F</td><td></td></tr>
<tr><td>Zone 2</td><td>-50° to -40° F</td><td></td></tr>
<tr><td>Zone 3</td><td>-40° to -30° F</td><td></td></tr>
<tr><td>Zone 4</td><td>-30° to -20° F</td><td></td></tr>
<tr><td>Zone 5</td><td>-20° to -10° F</td><td></td></tr>
<tr><td>Zone 6</td><td>-10° to 0° F</td><td></td></tr>
<tr><td>Zone 7</td><td>10° to 20° F</td><td></td></tr>
<tr><td>Zone 8</td><td>20° to 30° F</td><td></td></tr>
<tr><td>Zone 9</td><td>30° to 40° F</td><td></td></tr>
<tr><td>Zone 10</td><td>40° to 50° F</td><td></td></tr>
<tr><td>Zone 11</td><td>50° to 60° F</td><td></td></tr>
</table>

Hardiness zones in the United States

in order to survive year-round, while in Zone 6, in the Midwest and West, the mercury never goes below –10 degrees Fahrenheit. Plants in Zone 9 (Florida, southern Texas, southern Arizona, and California) never have to deal with temperatures lower than 30 degrees Fahrenheit. Many plants are rated by zone or hardiness level. To be safe, choose plants that are rated one zone below the one in which you live.

No matter where you live, microclimates can enable you to grow plants that do not normally live in the area, making a successful niche market for your produce. Microclimates can be created through the use of trees or other windbreaks, from natural depressions or rises in the land, or from man-made weather breaks, such as buildings. You can always create your own artificial microclimate through such means as greenhouses and indoor plant farming. We will discuss this more in chapter 5.

Equipment

Once you have assessed your land, consider the equipment needed for your various crop and livestock enterprises. Machinery and high-quality hand tools are expensive to buy and costly to maintain and repair.

Machinery is a tool to help you do more work faster than you could by hand. If you don't use the time saved to do more profitable things on your farm (like planning or marketing), your machinery is just an expense you don't really need. Machinery should enable you to produce the volume of business necessary to make a profit with as little labor as possible.

Remember that the farm has many free energy and power sources, starting with the sun, which can be utilized to reduce your machinery, labor, and resource needs. For example, crop rotations — using plants themselves to nourish the soil — improve your yields without requiring purchase of energy-consuming commercial fertilizers, and thus make your operation more sustainable.

In addition to machinery, equipment includes the physical plant of the property — buildings and fencing. With careful placement of

FREE ENERGY SOURCES

Sun

- Photosynthesis from the sun produces crops and forages.
- Sunshine generates heat, which may be stored by greenhouses or mulches.
- Sunshine may be used for water collection through condensation.
- Sunlight can sanitize soil, killing microorganisms such as bacteria and molds.
- The sun provides essential nutrients like vitamin D.

Water

- Running water may generate power.
- Rain (and snow) provides free water for crops and livestock. Collect it for future use.

Soil Enhancers

- Nitrogen fixation from leguminous plants eliminates adding nitrogen.
- Crop rotations provide elements to other crops, replacing applications of fertilizer.
- Organic matter from plants (green manure, pasture, crop waste) is fuel for microbial activity in your soil, which helps plants get the nutrients they need.
- Worms and various fungi/bacteria will migrate to and breed in healthy soil. These help aerate the land and provide essential minerals to the plants.
- Azotobacters are a form of free-living bacteria that can fix the nitrogen of the air into the soil. They will also be found in soil with a high organic content.

Energy from Livestock

- Livestock will naturally plow a pasture or garden as they graze (especially hogs).
- Manure from animals in either raw or compost form can be used to enrich the soil.

Other Free Energy Sources

- Wind may be used to generate power.
- Trees can cool livestock in their shade.
- Windbreaks prevent water and heat loss.
- Trees may provide a fuel or building source.
- Pasture and crop stubble can be used to feed animals.

pasture fencing, you can give different nutritional levels of forage to different-size animals or animals in different phases of production — like nursing cows or bred ewes. (It takes a higher level of nutrition for a female nursing and raising offspring than it does to maintain a pregnant animal.) For example, in general, legumes (clover, lespedeza, alfalfa) have a higher protein content and relative feed value (RFV) than grasses. To flush ewes (increasing ovulation), place them on lespedeza or alfalfa pasture. You can also get similar results with a 5- to 6-inch-tall grass pasture; even though the grass has a lower RFV, quantity can balance its lower quality. In the same vein, multiple-birth animals (sheep, hogs) benefit from weight gain from an increasing level of nutrition, which makes them ovulate more to produce more offspring. The right type of fencing allows animals to be easily moved to where they need to be according to their nutritional needs.

How do you determine your machinery needs and costs? Again, analyze your crop and livestock needs. We cover equipment and machinery in more detail in chapter 9.

The females of multiple-birth animals such as sheep and hogs should graze in areas with high-quality forage to satisfy the nutritional needs of pregnancy and lactation.

Getting Help

If you are new to farming, you may not be able to answer all of the questions that arise from your farm plan. By the time you finish reading this book, you will be able to answer more of them, but it is important for you to learn from a variety of sources. Go to your local library and do some research on enterprises that interest you. Talk to farmers you know, or go to a farmers' market in the area you want to move to and talk to the vendors there about what they grow and what works for them. Visit your local university extension office and get advice on crops and livestock that will work in your area. If they have a expert on direct marketing or adding value, plan on frequent consultations with him or her.

As you learn from all of these people, remember that everyone has answers, but they are not always the right answers for you or your farm — always get second, third, and fourth opinions. If someone says you can't do "X" on your farm, you will have to decide how important it is to you to do "X," and if it is important, research further to find out whether there is a way. I will speak more about resources for learning later on.

❖❖❖

FOOD FOR THOUGHT

Before you decide what to farm, read all that you can on as many different topics of interest as possible. Knowing where you are going is the first step to getting there. Starting a farm may seem like an impossible task, but if you have faith in yourself and your family and you do your research carefully, you will succeed in getting the farm — and lifestyle — you want.

❖❖❖

PART II

Farming

CHAPTER THREE

Some Principles of Good Farming

In addition to deciding where and what to farm, you'll need some basic guidelines on farming itself. Although how-to methods will be discussed later, you must learn basic principles to apply to your farm. We have already covered some principles. I would like to repeat them, just so you will be able to look at them as a group:

- Plan your farm and your goals carefully.
- Farm goals should be family goals.
- Look at the whole picture.
- A farmer should learn and grow through reading and meetings.
- Read, research, and experiment.
- Think.
- Old is not always obsolete. New is not always better.
- Natural fertility and slope of land are critical.
- To be sustainable, a farm must be profitable.
- A farm *must* be profitable.

Now, let's continue with some other important principles.

◄ *Farm goals should be family goals. Your children learn responsibility by being involved in the farm goals.*

PRINCIPLE: *To be sustainable, a farm must be environmentally sound and socially acceptable.*

For your farm to be sustainable, you must develop cropping and livestock systems that are environmentally sound and socially acceptable. For instance, plowing up and down the hill so your soil washes out onto a public road is not environmentally sound, because you lose a lot of healthy soil. Nor is it socially acceptable, because everybody pays for the cost of cleanup. As another example, huge feedlots of large numbers of concentrated animals are encountering more and more opposition today, some even from neighboring farmers. Lack of social acceptance will prevent this type of farm from being sustainable.

PRINCIPLE: *Avoid debt.*

Dealing with bankers is like selling through a middleman. The farmer makes less profit because he has to pay interest and principal. Start-up debt is okay if you can pay at least 20 percent (preferably 50 percent) on the land and stretch the payments out as low and as long as possible with no prepayment penalty when you have a good year. For all other projects, *grow* into the enterprise rather than *borrowing* into it. The principle is to avoid debt as much as you can and to make payments in line with farm production ups and downs.

PRINCIPLE: *Keep costs down.*

Whenever you try something new in the way of crops or livestock, do it on a small scale and grow into it while learning. It could save you lots of money.

PRINCIPLE: *Try for low inputs.*

Low inputs may improve your soil and make your operation more sustainable, in addition to saving money. Of course, the less money you spend, the more you have to work with.

PRINCIPLE: *Do things on time.*

Accomplishing tasks on time is an important principle of farming that requires your labor and machinery requirements to match. Fields are best sized to what you can do in one day. You can drive only one tractor or tiller at a time. Timing is important because of the seasonal and cyclical nature of farming. If you want your cows to calve close together, the bull must be with your cows for only 60 days, or two heat cycles. Close calving means more attention on your part, and a more uniform-sized calf crop for selling purposes.

PRINCIPLE: *Plan your farm to minimize your work.*

Work can be minimized by planning your farm layout wisely. Run a travel lane with access to crop fields and pastures down the middle of your farm, to allow for efficient movement of livestock and machinery; create square or rectangular fields to maximize the efficiency of your machinery; and plan placement of sheds and gardens for best access from the house and fields to save time and energy.

If you are raising herd-type animals (for example, cattle, sheep, elk), maintain several animals whenever possible. Herds of animals are more content than are one or two animals, who are always wanting to rejoin the main herd. Uniform bunches of animals enable you to creep-feed calves on higher-quality forage than is needed for the mother cows. Remember that culling is important.

PRINCIPLE: *Develop a system of production that balances farm resources and available labor.*

Small acreages lend themselves to good timing. Due to their small size, you can reach locations quickly. With low numbers of livestock, tasks do not take long to perform. This often allows you to beat the weather or use it to your advantage.

PRINCIPLE: *Keep good records.*

It's essential to keep careful financial and performance records and a diary. One of the most important things you need to know is the cost of production for each enterprise. It is nice to know ewe #29 had triplets for the last three years. It is nicer to know that it cost "X" dollars of feed for the ewe and "Y" dollars for the lambs, and the cost of the pasture, hay, equipment, and so on, allotted to her was so many cents per pound, because now you know what price you need to sell at to gain a profit.

PRINCIPLE: *Learn basic veterinary skills and tasks.*

Whether new to the farm or established, good how-to skills are important. Vet calls today are easily $25 just for the trip, plus labor and medicine, so it pays to learn how to dock sheep tails, castrate calves, and deliver babies, for example.

PRINCIPLE: *Learn carpentry, electrical, and machinery repair skills.*

Carpentry, electrical, and machinery skills all reduce costs. If you do not have these skills, some courses could be helpful.

PRINCIPLE: *Learn stockman skills, and keep gentle livestock.*

Stockman skills are also important, but you may not be able to learn them in school. Talk to someone in the business and see what he knows, or read books and magazines. A basic example is knowing where to stand when herding cattle — standing in a place in their field of view saves time in herding and reduces the animals' stress level.

Research has shown that gentle livestock reproduce and grow faster and better than do wild, nervous animals. But there are other reasons that gentle livestock are important. If you have a pet cow or ewe, for instance, one that will always come to a feed bucket when you call, that animal will help you move your livestock from one pasture to the next or into the working corral area with a minimum of trouble. The pet will also bring the herd in for a close visual inspection for new babies, bad eyes, sickness, or other problems. Being around your animals, patting and scratching the pets, and observing the others all help keep them gentle. Using a treat like cattle cubes or an ear of corn also speeds along the process.

PRINCIPLE: *Take good care of your buildings, machinery, and livestock.*

Buildings and machinery that are taken care of will always cost less to maintain and will rarely need replacement. Obey your engine care instructions on machinery, and frequently inspect your buildings and machinery for wear, weather damage, and general condition.

Livestock also respond to better treatment. Healthy, stress-free livestock will gain weight more quickly, stay in better condition, and have better performance in birthing and raising young. Learn everything you can to keep your livestock healthy and unstressed, and make sure they have adequate shelter, feed, and water.

PRINCIPLE: *Have a good water system, and save every drop of water that falls on your farm.*

A good water system for livestock is essential. Again, it is better if your livestock can go to the water, instead of you hauling it to them. I have found it tiring to dip water out of a pond and haul it to the hog water tank on another part of the farm. Your own pond or lake with a submersible pump and plastic pipe makes life more pleasant for you and your livestock.

Water can be stored in ponds and in the soil itself. For instance, if your soil is 4 to 5 percent organic matter, it can absorb 4 to 6 inches of rain per hour without erosion, which would cause runoff. If your organic matter is only about 2 percent, your soil can absorb only ½ to 1½ inches of rain per hour. (See chapter 5 for more on water.)

PRINCIPLE: *Maintain or improve the soil fertility.*

Maintaining soil fertility is a process of thinking about the "why" of the way you do things on your farm. Several processes improve and help fertility, like composting on either large or small scale, animal manures, and green manures for cover crops.

Animal manures can be good for the soil. One of the best ways to spread manure is to let the animals do it. Hauling piles of manure from animals standing in a muddy barn lot requires that you spend time and labor forking out the manure, plus fuel and expensive machinery (the manure spreader) to spread it. If you keep your animals on pasture instead, they will spread it across the field as they defecate while they graze. This natural method lets your animals do the work for you.

PRINCIPLE: *Let the animals do as much feed harvesting on their own as possible.*

It is much easier to stockpile fields of grass for winter feed and drive the livestock to that field one time than to haul hay daily. To put up hay for winter forage requires time, money, and equipment. You may not be able to avoid hay altogether, but the cost savings when you can are money in your pocket.

PRINCIPLE: *Use crop rotations.*

Crop rotations are vital to a sustainable farming system. They enhance soil fertility and help control plant diseases, weeds, and erosion. Soil fertility in a rotation is maintained by the growing of a legume (like clover) or sod crop (grass) to provide nitrogen fixing and buildup of humus. Crop rotations also lend themselves to livestock usage; the legumes and sod crops in the rotations may be used for grazing and for hay.

PRINCIPLE: *Have 2 years' worth of hay and grain in storage.*

This is a goal you should try to achieve for a number of reasons. The obvious one is weather. You never know when the winter will be longer, or the summer hotter and dryer, and you will need extra feed to supplement your livestock on pasture. If you are having a weather problem, chances are your neighbors are too, and feed prices will be high. When you have your own feed, you are more likely make it through to rain. If not, you will be selling your live-stock on a depressed market. Fortunately, droughts and bad winters do not happen too often, but it is always best to be prepared.

Here are some principles we will cover in future chapters:

- ❖ To be sustainable, a farm must be diversified.
- ❖ A farmer must be skilled at buying and marketing.
- ❖ Plan to have products to sell year-round.
- ❖ Plan to have produce available when others do not, or have unique products.

These will be covered in more detail in chapters 8 and 9, but the basic idea is to learn to market, and to have something to sell at all times. Your bills are monthly, not seasonal, so you should have something to sell year-round. A diversified farm is protected from weather conditions or price variabilities that hit one crop or animal but not another. Markets vary at different times: When hog prices are low, sheep prices may be high. You should also diversify your marketing methods, so you are not dependent on only one type of income. You also need to set and control your prices, rather than just taking the price someone offers.

And finally, some advice for you to practice on yourself:

- ❖ Be disciplined.
- ❖ Don't procrastinate.
- ❖ Practice scheduled, efficient, and productive work habits.
- ❖ Keep a positive attitude.
- ❖ Be happy.

❖❖❖

FOOD FOR THOUGHT

Learning the basic principles of farming will provide you with the tools needed to run a successful operation. Though all of the principles in this chapter are important, the most important ones are the last few mentioned. A positive, happy attitude and a disciplined approach to problem solving will make your goals easier to obtain.

❖❖❖

CHAPTER FOUR

A Living,
Healthy Soil

Most farmers of any type will say it takes 5 years to learn how to farm a specific piece of ground on a particular farm. Most business people will tell you the learning curve for a new start-up business is 10 to 12 years. To be organically certified, your land must not have had chemical fertilizers or weed and pest sprays used on it for 3 years or more, to give the soil a chance to renew itself and return to life.

All these statistics are based on the assumption that you can survive in your business long enough to reach a mature level. All businesses, farming included, need to be least-cost producers in order to survive the ups and downs of the marketplace. To work at least cost, to get the most out of your land, to be sustainable and profitable, you must have good soil.

It is best, of course, if your farm begins with good soil. If it does not, though, don't be discouraged. There are techniques that will increase your soil's fertility, including crop rotation, application of organic matter, and careful choice of crops. The 4-acre Student Garden in Santa Cruz, California, started as a hillside

≺ *A living, healthy soil is the foundation for your farm's productivity and success.*

51

with a clay soil that would barely grow weeds. Within 3 years, using organic methods, the students created an excellent fertile soil. Note, though, that improving soil is not a quick fix. You can count on a minimum of two years, and probably more, before your soil is what you want it to be.

> **PRINCIPLE:** *The foundation of your farm and your most important production tool is a living, healthy soil.*

Here we will go into some technical detail, because good soil is the foundation of a good farm — and to attain good soil, you must understand the basics of soil fertility. Even experienced farmers may not have a good grasp of soil mechanics, because they have only learned to add N-P-K (nitrogen, phosphorus, potassium) at the right time to make their corn and soybean rotation work. Good soil has a much greater complexity than that, an interaction among your plants and livestock, insects and worms, microorganisms, and soil types.

The Physical Nature of Soil

Soil is formed by water and wind erosion, and temperature variances, which crack large rocks into smaller rocks (in Missouri, sometimes not small enough), then down into particles that combine with decomposing organic matter (dead and decaying plants and animals).

Nature forms soil in layers. The top layer is the highest in organic matter and, hence, the most fertile. The top layer is usually 6 to 7 inches in depth. This layer is what is normally plowed for planting. The next layer is the subsoil and is obviously weathered. It has little organic matter. The third layer, the substratum, is more or less the parent rock/material from which the particles for soil are formed.

Soil Types

Because soils contain particles of varying size and materials, and organic content also varies in different areas, there is a wide range of soils. Three principle types are recognized: sandy, clay, and loam soils.

As mentioned earlier, there are more than 18,000 different soils in the United States. Every state (and every farm) contains many different soils. These soils are members of eighteen different orders of soil. Orders that cover a broad range of the United States include the ultisols, the mollisols, and the alfisols. *Ultisols* are a southeastern soil, highly weathered, with subsurface clay. The clay stores water and nutrients for plants to use. *Mollisols* are primarily a midwestern soil, with a fertile surface layer of high organic content. They are excellent for corn, soybeans, and wheat. *Alfisols* are productive soils of the mideastern states. They require careful management.

Soil Content

The orders and suborders define the soil types, but what you really need to know is what your soil contains. Soil types are classified according to content: sand, clay, loam, and organic matter.

Sandy Soils

Sandy soils are up to 70 percent sand by weight. They are known as "light" soils, or soils that drain readily, because sand particles are fairly large and irregular in shape, and do not press together tightly, thus allowing water to make its way among the particles. Sandy soils are usually lower in nutrient content than the other soil types, as the water washes the nutrients out. Organic matter can do wonders for sandy soils, but they still require special management, as we will discuss later. Sandy soils come in different classes, such as *sand* and *loamy sand*.

Clay Soils

Clay soils contain at least 35 percent clay, and usually 40 percent or above. Class names are *clay* and *sandy clay*. Clay soils are made up of flat particles that can pack tightly together, making for poor drainage and aeration. Clay soils are sticky when wet and hard when dry, and if cultivated at the wrong time will give you a summer of hard, lumpy soil to work with. If you plow clay soils when they are too wet, they will puddle or run together and be like brick when dry.

Loam Soils

Loam soils are a mixture of 45 percent sand, 40 percent silt, and 15 percent clay particles. When sand is the dominant particle, it is called *sandy loam*; likewise with silt loam and clay loams.

Loamy soils are what you want, because they combine the best characteristics of clay and sand soils. They have good aeration and drainage, but retain good water-holding capacity, which preserves nutrients and prevents nutrient leaching.

See the resource list in the appendix for contacts for aerial photos and topographic and soil maps for your area.

Working the Soil

Like animals, plants, and other living organisms, soils respond to good care and suffer under bad handling.

Too wet   Still too wet   Just right

Working soil when too wet destroys structure tilth and harms beneficial microbial life. Test your soil before you work it.

To test your soil to see whether it is of the right moisture to work with, make a loose ball of dirt and drop it from about waist high. If it breaks apart, your soil is ready to work. If it does not break, then wait a half day or a day with no rain before you start working the ground.

If you are working your ground with a tractor and there is water in the plow sole, *stop* — it is too wet. If the plow furrows glisten in the sun, or look slick rather than crumbly, the ground is too wet — stop plowing. When turning over the furrow slice, it should be crumbly and falling apart.

Our Living Soil

Living soil is much more than just the different kinds of mineral particles that hold the plants upright. It is teeming with life throughout its strata, from bacteria to worms. Most of your soil's organisms

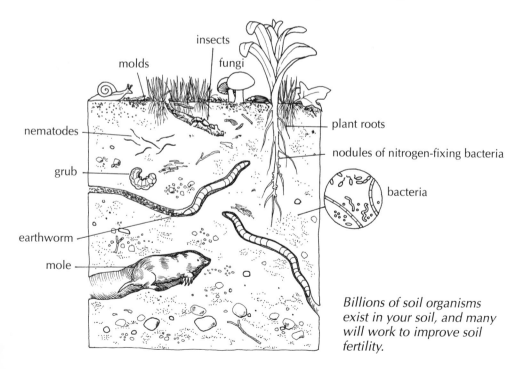

Billions of soil organisms exist in your soil, and many will work to improve soil fertility.

inhabit the top 6 or 7 inches, the topsoil. These organisms include insects and other "bugs," earthworms, bacteria, and fungi.

Earthworms

The earthworm is the largest of the soil's creatures, and performs a multitude of tasks. Burrowing lets air into the soil, as well as water. Worms move surface organic matter underground into their burrows. This is one way they fertilize your soil. The other way is a special talent of worms: When they eat organic matter and it passes through their body, the result is worm manure — worm cast-

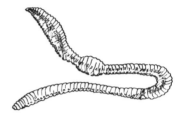

Night crawlers (Lumbricus terrestris) *are not native to the United States, but they have adapted to life in cultivated fields better than have native earthworms.*

ings — which contain 5 times more nitrogen, 7 times more phosphorus, 3 times more magnesium, 11 times more potassium, and 1½ times more calcium than the surrounding soil.

USDA researcher Henry Hopp conducted many extensive studies on worms in the 1940s, and noted that worms move the soil by two methods. In loose soils, they slip through the holes; in tight soils, they simply eat their way through. While worms deposit some of their castings on the surface, they mainly deposit them underground, which fertilizes your soil. In Australia and New Zealand, in areas where worms were not natural to the soils, pasture production increased 30 to 50 percent with the introduction of earthworms.

The best way to maintain a high adult worm population is with mulch or a cover crop. Worms are most active in the spring and fall and go into a kind of hibernation when it is too dry or too cold. A cover crop growing in the fall keeps the ground temperature from dropping quickly and killing your worms. The cover crop lets the soil freeze slowly, allowing the worms to move deeper and thus survive. The larger the adult worm population you can maintain, the more young you will have and the more fertility your soil will have.

Bacteria

The earthworm is one of the largest inhabitants of the soil community, but there are billions of organisms in each gram of healthy topsoil. You may have up to 2 tons of bacteria per acre. Bacteria are either aerobic (they need air to live; these are the most beneficial ones) or anaerobic (these do not need air to live).

Bacteria in your soil make nitrogen and sulfur compounds into a form usable by plants. This is important because nitrogen is often the limiting factor in plant growth. Soil may contain only small amounts of nitrogen in forms that plants can utilize, and some plants are heavy feeders of nitrogen. There are several bacteria that can make soil minerals available for plants.

Azotobacter is a free-living microorganism that fixes nitrogen from the soil and air into its body tissue. Found naturally in soils, azotobacters can be encouraged to multiply by the addition of organic matter, humus, and humic acids to the soil.

A nitrogen-fixing bacterium called *Rhizobium* occurs naturally on leguminous plants like peas, beans, and clover. This bacterium takes nitrogen from the air and converts it to a form that plants can utilize. The population of *Rhizobia* increases as the amount of organic matter and humus in the soil increases. It is also possible to buy *Rhizobia* in small packets from many companies in order to inoculate seeds. This may be necessary if you have no *Rhizobia* in your soil, which is possible if no nitrogen-fixing crops have ever grown there. We'll discuss nitrogen fixing later in this chapter.

Another important bacterium is the *actinomycete*. Even though actinomycetes are bacteria, they also share some similarities with fungi. They act as decomposers of organic matter and facilitators in humus formation. Actinomycetes can be found everywhere, but are particularly rich in sod soils and compost. To be most effective, they need a neutral pH of 6.0 to 7.5.

Fungi

These organisms include mold, yeasts, and mushrooms. They get their nutrients from decomposing organic material or live organic material. Although some fungi are harmful, what we are concerned with here are beneficial fungi, such as *mycorrhiza*, a mycelium fungus that has a mutually beneficial relationship with plant roots. Plant roots provide carbohydrates to sustain mycorrhiza, and the fungus helps the plant absorb nutrients and water. About 80 percent of the agricultural plants grown have a mycorrhizal root association. Mycorrhiza can be enhanced with increased organic matter and minimal disturbance of the soil.

Soil and Plant Roots

As we have seen, your soil is both a physical structure and a group of living organisms. Like livestock, your soil needs food, water, and air to reproduce and function.

Plant roots help supply your living soil's needs, and connect the plant to it. They remove nutrients and water from the soil to feed the plant, but they also feed the soil. Plant roots are always getting rid of dead tissue, which makes excellent food for soil

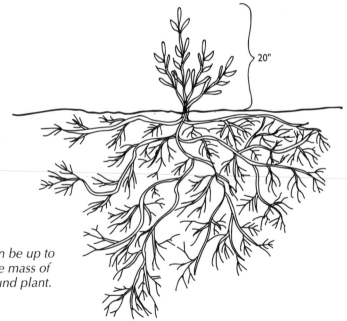

20"

Plant roots can be up to three times the mass of the aboveground plant.

microorganisms. Many food transactions occur right around the plant roots; this area is known as the *rhizosphere.*

The roots of plants are typically two to three times the mass of the aboveground plant yield and cover a large area. H. Dittmer, writing in the *American Journal of Botany,* 1937–38, found that a single rye plant had a root length of 377 miles. The root hairs numbered 14.5 billion. The surface area was more than 1/10 of an acre. Combined, the roots and root hairs had a length of 6,990 miles, with a combined surface area of 63,784 square feet — close to 1.5 acres.

As already mentioned, most of your soil's organisms are found in the upper 6 or 7 inches — the plow layer, or topsoil. Although most of the plant's roots are in this layer, many plants send roots 5 or 6 feet deep in search of nutrients. These deep roots help provide a structure to the soil, preventing erosion and increasing the soil's capacity to hold water. To achieve deep roots, the soil must contain good amounts of nitrogen.

A loose soil with good tilth or structure lets air reach to the roots, which are then able to work better. Tight, waterlogged soils have no air, and plants in them soon yellow and die.

Soil is a loosely connected body of particles of irregular shapes and sizes. The spaces between these particles form cavities, some large and some too small to be seen by the naked eye. Small spaces allow for capillary action by the soil, similar to a paper towel soaking up water. The various cavities, or pores, are filled with water and air, both of which are essential to plant growth. We will talk more about water in chapter 5.

Continuous cropping results in rapid loss of soil organic matter. Alternating shallow- and deep-rooted plants like clover and alfalfa provides better drainage from channels left by dead roots. Deep-rooted plants bring up minerals from the subsoil. A rotation with a sod crop maintains the organic matter supply and furnishes the raw food for soil bacteria. By keeping a crop on the land, most of the toxic nutrient leaching is minimized. Crop rotation improves yields as well as crop quality.

Crop Rotation

Crop rotation is the process of planting a different crop after each previous crop, which allows the different plants to take advantage of nutrients the previous plants didn't use, and to put different nutrients into the soil to avoid depletion of overall nutrients. For example, corn uses nitrogen; soybeans replace it. Following corn with soybeans avoids nitrogen depletion.

Crop rotation dates back to Roman times. Farmers in those days began rotations to replenish the land instead of using up its fertility and abandoning the field.

A rotation should be planned so that you have the greatest possible value of salable or usable crop material during a period of years. When planning a rotation, also consider labor needs and soil fertility. Ideally, a rotation will help spread out your labor needs, because you have a diversity of crops ready to harvest at different times of the year.

Crops are divided into three classes: grain crops, like wheat and barley; grass crops, including sods and legumes used for pasture/hay; and cultivated crops, like corn and soybeans. You can substitute crops within types, depending on weather conditions and year. For instance, you can plant barley instead of wheat; both are small grains. Ideally, farm fields would be square, but this often does not happen in real life, so just try to keep plots all the same size — that is, if you plant 10 acres of corn, follow it with a full 10 acres of soybeans, then 10 acres of hay, and so on.

Advantages of Rotation

There are many advantages to crop rotation. The biggest one is building and maintaining organic matter and putting nitrogen back in your soil by plowing under immature forage crops, primarily legumes. (Legumes are plants that fix nitrogen in the soil.) Crop rotation lets you grow a soil crop or legume crop on all the fields of your farm.

Here are some other advantages:

❖ Rotations of different crops provide varying root systems, some deep, some shallow, which bring different crop nutrients to the plow layer for use by the next planting.

❖ Rotations improve drainage tilth and water-holding capacity of the soil, which also reduces erosion.

❖ By alternating crops on the same fields, you use a natural method of breaking up insect pest and disease cycles, and this also helps in eliminating weed species.

❖ Rotation hosts beneficial microbiological life, which discourages disease; improves microbiological activity, which helps plants absorb nutrients better; and creates an inhospitable soil environment for many soilborne diseases.

❖ By rotating crops, a farmer's labor load is spread throughout the season, making for more timely operations.

❖ Rotation or diversification of your crops provides protection against total crop and economic failure, and provides year-round distribution of labor.

❖ Rotations cut costs and time by reducing purchased fertilizer and allowing easier tillage because of improved tilth.

❖ On the average, rotating your crops will give you a 10 percent increase in yields; continual planting of the same crop results in mineral depletion. Crop rotation allows you to make money while you build your soil.

Short-Term Rotation

Rotation can be either short term or long term. A short-term rotation will take place in 1 year or less. An example is fall-planted wheat overseeded in spring with red clover. To protect the wheat, the clover should be overseeded on a snow cover or on frozen ground, preferably both. This can be done by hand, by powered seeders mounted on tractors, pickup trucks, or all-terrain vehicles, and even by plane! After the grain is harvested in July, the clover grows through the wheat stubble and can be grazed in the fall or incorporated by plowing or disking before another fall grain crop is planted.

Another short-term rotation might be rye and hairy vetch seeded in standing corn. After the corn is picked, the rye and vetch grow through the winter and are then plowed under in the spring for a green-manure crop, after which corn, beans, or milo may be planted. A short rotation for vegetables might be broccoli followed by buckwheat, then perhaps rye or turnips.

Minor rotations are another type of short-term rotation. (Both major and minor rotations can be used on the same farm.) A minor rotation is when a small area is set aside for a short-term use — a truck garden, a hog pasture, a lambing area, for example. The minor rotation can be part of the major rotation, or it can be temporarily or permanently set aside and fenced. A more permanent setup of a minor rotation allows you to add feeding and watering systems.

An example of a minor rotation might be hog pasture, potatoes, then truck garden or alfalfa. The hogs turn up the soil for potato planting and contribute manure to the soil fertility, as does a legume like alfalfa. A late summer/fall garden of crops such as beets, cabbage, collards, lettuce, and spinach will allow you some late sales, along with extra greens at the table.

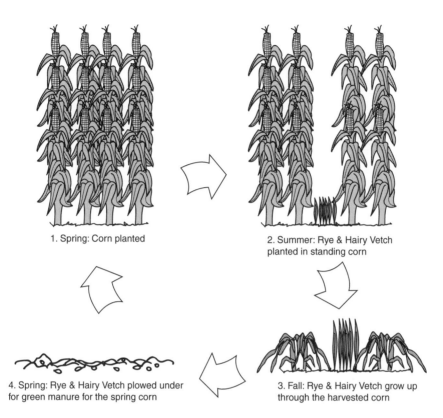

1. Spring: Corn planted

2. Summer: Rye & Hairy Vetch planted in standing corn

4. Spring: Rye & Hairy Vetch plowed under for green manure for the spring corn

3. Fall: Rye & Hairy Vetch grow up through the harvested corn

In this example of short-term crop rotation, corn is followed by rye and hairy vetch, which are then plowed under before corn is planted again.

Early Spring

Spring & Summer

Fall

Minor crop rotation in this example includes hogs in early spring, potatoes in spring and summer, and vegetables and alfalfa in fall.

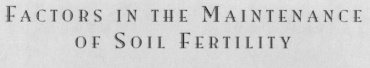

Factors in the Maintenance of Soil Fertility

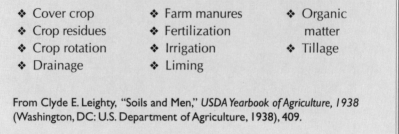

- ❖ Cover crop
- ❖ Crop residues
- ❖ Crop rotation
- ❖ Drainage
- ❖ Farm manures
- ❖ Fertilization
- ❖ Irrigation
- ❖ Liming
- ❖ Organic matter
- ❖ Tillage

From Clyde E. Leighty, "Soils and Men," *USDA Yearbook of Agriculture, 1938* (Washington, DC: U.S. Department of Agriculture, 1938), 409.

Long-Term Crop Rotations

Long-term rotations are usually 2 to 5 years. They are more complicated than short-term rotations, but usually not more difficult. They often include hay and sod crops, which are utilized by livestock. A popular eighteenth- and nineteenth-century rotation was the Norfolk rotation: root crop/barley/legume/wheat. The legume allowed introduction of livestock such as sheep or cattle into the rotation, by using the legume for grazing or hay.

Some common 3-year rotations are corn/rye/clover or corn/barley/clover. An example of a 5-year rotation is:

Year	Crop	Comment(s)
1	Corn	The corn crop allows weed control through cultivation; fertility for future crops is supplied by manure from cattle or sheep grazing the stalks after harvest.
2	Oats	Sow on the disked cornstalks.
3	Wheat	Makes a good nurse crop for the following clover.
4–5	Clover and grasses	Pasture for livestock grazing comes from the clover and grass, but you may also have some permanent pasture on land that cannot be plowed because of slope and erosion.

Note: With this rotation, you must plow only twice in the 5 years (corn and wheat), which helps control erosion.

A common 5-year rotation in the southern United States is given below (left), with a variation at right.

Year	Crop	Year	Crop
1	Corn	1	Corn
2	Soybeans or cowpeas	2	Cowpeas or soybeans
3	Wheat	3	Wheat, followed by cowpeas
4–5	Clover and timothy grass	4	Wheat
		5	Clover and timothy grass

Planning a Rotation

Every farm has its own set of management and climatic constraints to deal with, but there are some basic rules of thumb to rotations, as mentioned by Nicolas Lampkin in *Organic Farming*. The following ideas are adapted from his rotation designs for England, but are applied to circumstances in the United States:

❖ Alternate deep-rooted plants (e.g., corn) with shallow-rooted plants (e.g., cabbage) to improve soil structure and thus drainage.

❖ Alternate between plants having a high biomass of roots (legumes such as red clover and orchard grass) with crops with a low biomass root system, like corn and soybeans.

❖ Nitrogen-fixing crops like soybeans should be followed by nitrogen-using crops like corn.

❖ Keep the soil covered with crops as much of the time as possible to prevent erosion and reduce weeds (see box).

❖ Some crops grow extremely fast, like sun hemp, buckwheat, radishes, and corn. They should be alternated with crops that grow slowly, such as winter wheat and red clover. Slow-growing crops are more susceptible to weed pressure and should follow weed-suppressing crops like winter rye, which has an *alleopathic effect,* that is, an ability to suppress weed germination and growth.

❖ Alternate from leaf to straw crops to help with weed suppression. Mechanical cultivation reduces weeds in row crops, while straw crops shade out and steal nutrients and water from weeds, thereby stunting the weeds' growth.

❖ Alternate between fall and spring plantings of crops. This spreads out your workload, reduces weather risk, and helps suppress weeds that germinate at different times of the year.

❖ Balance your rotation between the cash and non–immediate cash crops, because, as I've said before, it has to be profitable to be sustainable. This system makes it more profitable because it spreads risk and allows for cash crops (corn, wheat) and crops that build the soil (legumes, grasses, cover crops).

Cover Crops and Green Manure

Cover crops are used to prevent erosion, shade out weeds, and protect the soil from freezing and thawing. They also may reduce use of herbicides and pesticides by providing a haven for beneficial insects, as well as breaking the disease cycles of a monoculture system. By improving the health and microbiological activity of the soil, they also improve crop yield.

Cover crops are not necessarily incorporated into the soil, although they may be used as green manure (see below). Using a legume like red clover or hairy vetch in your field-crop or market-garden rotations gives you a plant useful for animal grazing or as hay, plus the benefits to the rotation of soil cover and nitrogen fixation. Red clover provides 2 to 3 tons of dry matter (the weight of forage after drying) and 70 to 150 pounds of nitrogen per acre. Hairy vetch supplies 3 to 6 tons of dry matter and 40 to 150 pounds of nitrogen.

Green Manure

A green-manure crop is a cover crop that is incorporated into the soil, even if it was not planted for that purpose, such as weeds growing on a flooded bottom field. Green-manure crops are turned under for the purpose of adding organic matter and/or nitrogen to the soil. This reduces your fertilizer costs, as the cover crop will furnish enough nitrogen for the following crop.

For every foot of green-manure crop you turn under, you put about 1 ton of organic matter back into your soil. This is because your green manure is just as big below the ground (roots) as it is above. About half — 1,000 pounds — of this tonnage is lost almost immediately through evaporation. Thus, green manuring is used to maintain rather than increase organic matter.

Legumes or Non-Legumes?

Green-manure crops and cover crops can be legumes (see next section) or non-legumes. As green manure, legumes add organic matter and nitrogen. Non-legumes add only organic matter. Volume or bulk is the most important goal in supplying organic matter to the soil and is more easily achieved by crops such as rye. If the crop is too mature when turned under, much of the available nitrogen may be used up in allowing bacteria to decay the crop. But with legumes, there is enough nitrogen to decay the crop and have nitrogen available for the next crop.

Nitrogen and Legumes

Nutrients move slowly in the soil. Roots usually grow toward the nutrients when they are activated by nitrogen and phosphorus (nutrients) already present in the soil. This makes nitrogen a necessary component of plant growth.

Legumes are crops, such as soybeans and hairy vetch, that "fix" nitrogen (that is, move nitrogen from the air or the soil to storage nodules in the plant), allowing increased plant growth and deeper roots. They have a *Rhizobia*-bacteria connection that allows legumes to take nitrogen from the air and store it in nodules on the plant root. This nitrogen is then released when the crop is plowed under and decomposes. You may need to supply the initial *Rhizobia* by applying an inoculant consisting of *Rhizobia* bacteria to the seeds to start the process.

The nitrogen in the legumes you turn under comes from both the nitrogen they drew up from the soil and the nitrogen they fixed from the air. Roughly one-third of the nitrogen comes from the soil and two-thirds comes from the air. If you sow a combination, say a (non-legume) rye and hairy vetch (a legume), the rye helps capture what nitrogen the vetch missed in the soil and uses it to grow lots of organic matter. While the rye uses the free nitrogen, the vetch

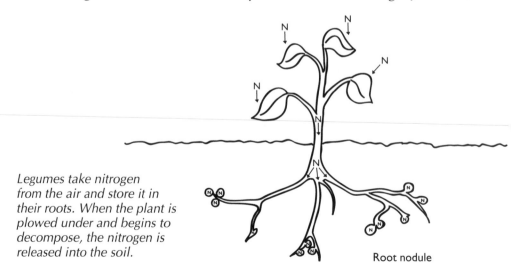

Legumes take nitrogen from the air and store it in their roots. When the plant is plowed under and begins to decompose, the nitrogen is released into the soil.

Root nodule

continues to fix more nitrogen from the soil into a plant-usable form. The rye will hold the nitrogen inside its slender green stems until both the leguminous and non-leguminous plants are turned under, furnishing enough nitrogen for the next crop.

The amount of nitrogen is determined by the time of cutting — the more time a plant has to gather nitrogen in an unstressed growth period, the greater the yield. Dr. C. J. Willard, an Ohio expert on legumes, noted that sweet clover in the Midwest cut the final week of April contained 124 pounds of nitrogen, while that turned under the last week in May contained 160 pounds.

The best time to incorporate (plow under) a legume is in early bloom, when it is most succulent. Controlling the growth stages of legumes and non-legumes to keep them consistent can be accomplished by mowing or grazing.

In southern areas of the United States, where a green manure is plowed under, it may have to be followed by a fall crop like rye to hold the nutrients until the new crop is ready to be planted in spring. This is because milder winters promote continued soil activity, which uses up the available nutrients. Also, many regions of the South suffer more soil leaching, allowing rain to drain away the necessary nutrients. The rye will prevent this leaching and reduce erosion.

NITROGEN FIXATION

1. Use the correct rhizobial inoculant for the legume you are growing. Ask your local seed dealer for the correct inoculant.
2. Soils need iron, sulfur, and molybdenum for nitrogen fixation to function properly.
3. A green-manure crop plowed under causes a large increase in microbiological activity. Soil bacteria reproduce themselves and double their population in as few as 7 days.
4. Weather affects the nitrogen release from green-manure crops. Warmer temperatures and field capacity of soil water at approximately 60 percent work the best for nitrogen release.
5. Most soil bacteria need a pH between 6.0 and 8.0 to perform well.

Best Cover Crops

Leguminous cover crops take nitrogen from the air and put it in the soil. They will also often attract beneficial insects.

Clover, Berseem *(Trifolium alexandrinum)*
Clover, crimson *(Trifolium incarnatum)*
Clover, red *(Trifolium pratense)*
Clover, subterranean *(T. subterraneum)*
Clover, sweet (white and yellow) *(Melilotus alba;*
   M. officinalis)
Clover, white *(Trifolium repens)*
   Ladino (*T. repens* forma *lodigense)*
Cowpeas *(Vigna unguiculata)*
Field peas *(Pisum sativum* var. *arvense)*
Medic *(Medicago)*
Vetch, hairy *(Vicia villosa)*
Vetch, woolly-pod *(Vicia dasycarpa)*

Here are some non-leguminous crops that are compatible with legumes. These furnish large amounts of biomass and increase nutrient-holding capacity when planted as a mix with legumes.

Barley *(Hordeum)*
Buckwheat *(Fagopyrum esculentum)*
Oats *(Avena)*
Rye, winter *(Secale cereale)*
Ryegrass, annual *(Lolium multiflorum)*
Sorghum/Sudangrass hybrids *(Sorghum bicolor* x *S.*
   bicolor var. *sudanese)*
Wheat *(Triticum)*

Feeding Livestock in Rotations

A crop rotation with legumes and non-legumes provides your farm with the opportunity to increase organic matter and soil fertility. The main purpose of pasture is for livestock feed — either by grazing animals directly on the pasture or by cutting and storing it for later feeding. Some of your fields will probably be in permanent pasture and some will be mowed for hay.

Mowing or grazing a plant causes some roots to die. Roots continually grow and die throughout the season. As stated earlier, the root tonnage per acre often exceeds the plant yield threefold, returning organic matter back to the soil.

The time needed for a pasture to recover after mowing or grazing depends on several factors, like fertility, soil, and moisture, but mostly on the degree of defoliation (what percentage of the plant was removed). The higher the percent removed, the longer the recovery or regrowth period. Mowing removes plant growth uniformly; grazing does not.

Mowing

If pasture grass reaches 12 inches or taller, it should be harvested for hay. At early bloom, non-legume plants have reached their maximum growth. For legumes, the point where they have the most nitrogen is early bloom. This also produces the highest-quality hay, if you intend to mow it. With legumes, yield goes up but quality goes down after this stage of growth.

Grazing

Jim Gerrish, an intensive-grazing expert at the University of Missouri, says that in general, animals should be turned into pastures when grasses are 6 to 10 inches in height and grazed to about half the current height. This avoids cropping the lower, growing stage of the grass. In other words, "Graze half and leave half."

Management-Intensive Grazing

There are several types of grazing methods you can use, but management-intensive grazing seems to be the best method now available. Basically, management-intensive grazing is having a minimum of sixteen grazing paddocks on which cattle are moved every one or two days, with each paddock receiving as long a rest period as possible.

In this system, grass is a crop and cattle, sheep, or goats are your combines. You "harvest" your grass when ready, just as you would corn or beans. With a standard rotation, you use only 35 to 40 percent of the forage available. Management-intensive grazing uses 80 to 85 percent of the available forage.

In New Zealand, farmers using this system are getting 700 pounds of beef per acre without irrigation or fertilization. On irrigated alfalfa, close to 2,000 pounds of beef per acre has been attained. The upper limits of this method of grazing will be determined by forage varieties, weather, and your management, combined with cattle genetics.

A good rule of thumb is not to have any cattle on any paddock more than 4 days. You may have to shift your livestock at different rates, depending on the time of year. In Missouri, fescue needs 60 to 90 days rest in July and August, but 30 days would be too much in May, when grass is growing rapidly.

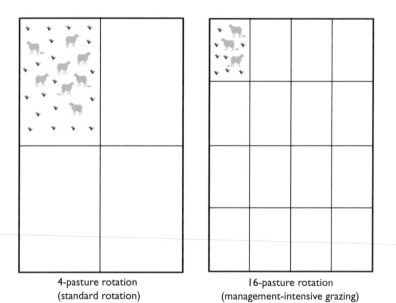

4-pasture rotation
(standard rotation)

16-pasture rotation
(management-intensive grazing)

In a standard four-pasture rotation, the forage is used 25 percent of the time and rested 75 percent of the time. In a sixteen-paddock management-intensive grazing system, the forage is used only 6 percent of the time and rested 94 percent of the time.

Pastures for Erosion Control

Pastures will do more than provide food for your livestock. Sod pastures, which provide a thick, even sod covering with the grasses interlocked, are important in controlling erosion. Kentucky bluegrass and perennial ryegrass grow into sod pastures. In contrast, many forages (e.g., alfalfa and warm-season grasses) grow in bunches. The better the sod, the more erosion is controlled — which keeps the soil and nutrients you want to have on your land.

Weaver and Noll, in 1935, tested an upland silt loam soil with a 10 percent slope and grass 2 inches high. Under those conditions, excellent sod in a 1-inch rain in 30 minutes had only 12 percent runoff and no erosion. Fair pasture had 55 percent runoff with some erosion. Poor pasture had 75 percent runoff — a soil loss of 4.6 tons per acre, with the only difference being the sod under the grass!

In *Quality Pasture,* Allan Nation notes, "USDA studies in 1930 showed that rotating cropland through sod-farming pastures for three to five years increased usable rainfall in Iowa by 6.4 inches a year over continuous corn. Grass crops absorb 87.4 percent of the rainfall versus 69.4 percent for a field of corn." In other words, good pasture increases the amount of water you can save for your farm — which leads us to the next chapter.

❖❖❖

FOOD FOR THOUGHT

The soil is the foundation for your farm. Good soil allows you to create an abundance of nutritious crops and healthy livestock. Take care when choosing crops and livestock, and apply management practices that will allow your farm to maintain or increase its soil fertility.

❖❖❖

CHAPTER FIVE

Weatherproofing Your Farm

The climate and weather of your particular region are natural resources. Although we seldom think about it, our soils and their fertility are a direct result of climate, because climate determines which plants will grow and which animals will inhabit the area. Weathering of rock, combined with the dead plants and animals, makes up the nutritional basis or fertility of your soil.

The Effects of Climate

It is important to match the climate with the type of farming you wish to do. Although you can grow just about anything you want wherever and whenever you want, it may not be cost-effective. For instance, growing catfish in Alaska would be too expensive because water temperature needs to be 65 degrees Fahrenheit for the catfish to begin eating and at least 80 degrees for quick, efficient growth. The cost of constantly heating a catfish tank or pond would be prohibitive.

≺ *Planting a shelter of trees for your farm will provide microclimates that enable your crops and livestock to be more productive and protected from weather extremes.*

Look for crops and livestock that thrive in the climate and weather of your region. Consider the crop's or livestock's home climate. If a breed of cattle or sheep has been raised in an arid region of the country, it may not do well if it is raised in a humid, wet area, which might cause parasite problems. Consider what is already being raised in your area. If you are new to farming, ask older farmers in the region what crops they grow now and what crops used to be grown in the area.

Temperature

Varying temperatures determine what plants grow best in what region, inspiring nicknames such as the Corn Belt and the Cotton Belt.

Low temperatures are not necessarily bad; many plants, bulbs, and seeds need a period of exposure to cold in order for them to germinate and grow properly. Snow is an excellent insulator and is very important in producing a good wheat crop.

Plants' life cycles are tied to the seasons and to temperature changes. Annual plants live only 1 year, producing seed, then dying when the temperature drops. Perennials die back to the ground after a hard freeze and survive over winter to grow again next year.

Water

Water (or the lack of it) affects everything on your farm — you, your livestock, your plants, and your soil. Living creatures need water for digestion and secretion, to regulate body temperature and feed tissues, to serve as a lubricant for joints and muscles, and to protect embryos in the womb.

Water and Livestock

An embryo is 90 percent water. A newborn calf is 75 to 80 percent water. An adult cow is 50 to 60 percent water. Animals can lose all their fat and half their protein and still survive, but

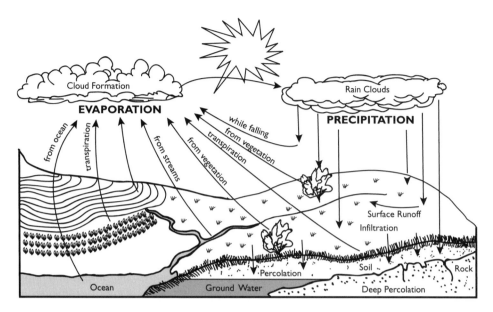

Catch every drop of water that falls on your farm through cropping and tillage methods that slow down evaporation and water flow.

the loss of one-tenth of their body's water means death. You can quickly see how important water is to your farm.

Animals may need large quantities of water. An adult human normally requires 5 to 6 pints of water daily. A mature cow requires 50 to 90 gallons per day. This does not mean you need a lake for each animal. Adequate forage reduces the need for freestanding water. Sheep can go weeks without drinking and may thrive without freestanding water if they have good pasture. The nutrient value of crops is also affected by the amount of moisture available to them.

Water and Plants

If soil is low in water, plants will find it difficult to grow or to perform all the functions that water facilitates, like transfer of nutrients through roots to other plant parts. Minerals in the soil can be absorbed by plants only when the minerals are in solution. Water is necessary for photosynthesis, the process by which plants extract energy from the sun. Without water there would be no microorganisms in the soil. Some of these organisms are good, like teramycin (used as antibiotics), and some are bad, like apple scab fungus, whose spores are loosened by large raindrops, but they all require water.

Water Needs Vary

The amount of rainfall is a major factor in determining which plants grow best in your region. Xerophytes are plants that can grow with limited water. Xerophytes may be succulent or nonsucculent. Succulents are plants such as cactus and milkweed that contain water storage cells to sustain them.

Western regions of the United States are known for their grass production, as many grasses can tolerate low natural rainfall. Gama grasses, curly mesquite, saltbush, and shrub oak all have xerographic (arid-tolerant) modifications. These plants grow quickly in the spring or following a rainy season. The shoots mature and then die, but the roots may survive for a year with no rain. The dried tops will furnish food for livestock. In southwestern Texas, sheep and cattle are grazed year-round on these grasses, which cure on the stem. We cannot change the weather, but we *can* take advantage of it.

Like most plants, corn needs water most when it is growing rapidly, such as when the silk is being fertilized to form the corn seed (kernels).

Hydrophytes, in contrast to xerophytes, are plants that grow in or near water, such as water lilies. Mesophytes fit between the two extremes, and they include most crops commonly grown. They also include both warm-season grasses (big bluestem, for example) and cool-season grasses, such as fescue.

Plants lose immense amounts of water through transpiration (evaporation). The rate of transpiration increases with temperature and light intensity. In addition, as plants grow they gain more leaf surface, which increases their rate of transpiration. Actively growing plants contain more water than they do solids. For every pound of plant dry matter, there are 5 to 10 pounds of water — but to produce that pound of dry matter requires several hundred pounds of water. The most efficient plants (pineapple, cactus, milo, gama grass, for example) use 250 to 300 pounds of water to produce 1 pound of dry matter. Less efficient plants (pumpkin, squash, cucumber, for example) may use 1,000 pounds of water to produce 1 pound of dry matter. Corn may transpire up to 2 quarts of water per plant per day, or 300,000 gallons of water per acre. Aquatic plants merely need to absorb the water — about 10 pounds per pound of growth. Which plants you choose will depend on your climate.

CLIMATE AND LIVESTOCK

Climate and weather affect livestock both directly and indirectly:

1. The climate can cause species to evolve with specific adaptations to that temperature. Some breeds of cattle, like Santa Gertrudis, were developed for hot-weather areas and have shaker muscles that help them get rid of flies, which are more numerous in hot areas. Some woolly animals, like alpacas, will find it difficult to cope with high temperatures, but do well in colder regions.

2. Indirectly, weather affects the vegetative growth in a region, which in turn makes some breeds more adaptive to that region. Irish Dexter cattle, for instance, were developed in mountainous areas having short growing seasons and sparse vegetation; they are a small animal with a dual purpose for meat and milk. Just as there is an inherent fertility in soils, livestock has an inherent capacity to produce the best under whatever are their optimal climate conditions.

Most plants need water when their growth is rapid. For instance, corn needs water when the corn is silking the ear and being fertilized by the tassel (the male part of the plant) to form the corn seed (the kernels). Insufficient water at this time will cause ears to fill unevenly with kernels. Tree fruits and berries are another example: their water requirements are highest just before maturity, when they increase greatly in size. This means that you must keep track not only of how much rain falls in your area, but also of when it falls.

Climate and Parasites

Parasites and diseases thrive in warm, humid weather. Tomato blight, for instance, can attack tomatoes if the weather is hot and humid, causing leaves to yellow and fall off, reducing the plant to survival mode or death — and this leads to a decrease in production.

Wheat harvest in the Midwest is associated with the hot, dry days of July and August, but if the weather is hot and humid, a disease called rust may cut yield in half. The wind can carry rust spores from mature wheat in Canada all the way to Texas, affecting plants along the way. When it rains, spores are deposited on newly sprouted small grains.

Weather and climate also influence the degree of parasites in livestock. Roundworms, for instance, need moisture and rainfall for the larvae to be able to infect hogs and turkeys.

Rain is the main cause of worm problems in sheep and goats, particularly when combined with warm temperatures. Worms are not as much of a problem in the drier western regions and the cold northern regions as they are in the South. There, the warm, humid weather allows worm parasites in sheep to flourish.

Altering Your Farm Environment

So far, we have looked at principles of weather and climate and how they affect our crops and livestock and the choices we might make regarding what we will raise. We know that we cannot control the weather, so it is important to find ways to adapt to our conditions by protecting against the elements and conserving water. Cooperating with seasonal weather cycles will also reduce the need for purchased feedstuffs and eliminate birthing in bad weather. For example, I plan all my calves to birth in April and May, when the weather is settling and the grass is growing.

Providing Shelter

Animals on pasture will use shelter only when needed, but it is good to have in rough weather — particularly for young livestock.

If you are on a small acreage and have no access to land with previously built shelters or naturally occurring livestock protection, you will have to build some sort of shelter. In my opinion, the best animal shelters are three-sided, open to the south, well bedded with straw, and portable. Some people even build shelters from hay bales. Smaller, portable shelters allow for multiple uses for crops and livestock, and will change as your farming operation evolves. A shelter under 40 inches high is best for small livestock — sheep, hogs, and poultry — to prevent rain and snow blow-in.

Multiuse barns should be avoided. They are harder to adapt, are fixed in place, and with multiple livestock types in one building, can increase disease problems. If your building cannot be portable or you already have one in place, and you use the building for birthing animals, do not use it again for birthing for 6 months or more. This will help to prevent disease.

Different animals need different kinds of shelter at different times of production. For instance, a thick cedar grove is excellent for calving or lambing.

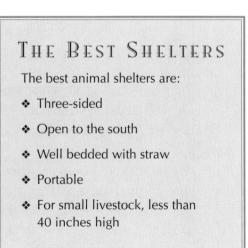

THE BEST SHELTERS

The best animal shelters are:

❖ Three-sided

❖ Open to the south

❖ Well bedded with straw

❖ Portable

❖ For small livestock, less than 40 inches high

The best animal shelters are portable, three-sided, open to the south, and well bedded with straw.

Cedars have vast amounts of fallen needles for bedding, and usually do not have low-hanging branches in a thick stand. They shed snow and rain in all but the heaviest downpours, and sheep like to be underneath them. I rotate pastures so that when it is time to lamb, the pasture with my cedar grove has grass that has not been grazed to the ground when I need it for nursing mothers.

Most livestock are pretty good mothers and do better on their own, rather than being confined to a small farm lot with a high potential for disease. They do especially well birthing on grass pastures when the grass is really growing (April/May for the Midwest) or whenever the temperature is in the 60s in your area. Sunlight and fresh air go a long way toward keeping your livestock healthy.

Predator-Proof Poultry Shelters

If you have predator problems, a fenced-in shelter to which you lead the livestock at night is essential. Poultry, in particular, need a place where they will be protected from coyotes, owls, raccoons, opossums, cats, and other predators. This may be a large poultry house to which chickens are returned each night, or a small portable coop such as a Smedley unit (normally for hogs).

A popular idea today is the chicken tractor, where poultry are kept in a portable 10-by-12-foot cage covered with chicken wire, with half of it covered in aluminum roofing. These open-bottomed shelters allow you to move your chickens daily through your garden or fields to eat insects, forage for weeds, and incorporate their droppings in the soil without damaging your plants.

Sheds

You will also need shelter for machinery. If your machinery is left out in the rain and sun, you are just throwing your money away. Rust and weather wear shorten the life of tools more quickly than does anything else. A machine shed where you can keep your tools and machinery dry, store repair equipment, and have a place to work on your tools will enhance your machinery, save you money, and improve your quality of life.

Finally, I also have a shed with a raised floor where I store my feeds, conveniently located near my livestock. It is carefully sealed to keep out rodents — and it does so most of the time.

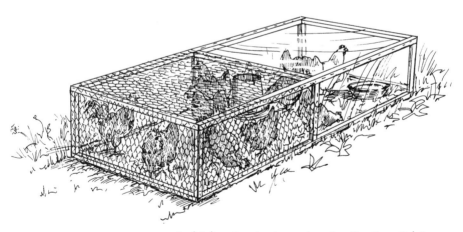

A chicken tractor is a win-win situation. It lets you move your chickens to a different spot in your gardens or fields every day, where they can devour insects and weeds and add their droppings to the soil.

Conserving Water

There are two ways to provide water for your farm. The first is irrigation (drip or standard), which waters crops when they need it. Irrigation facilitates practices such as foliar feeding (spraying a liquid fertilizer on the plant leaves) and incorporating fertilizer into the water. The drawback is the expense — the cost of irrigation equipment and the cost of the water. To provide water 12 inches deep on 1 acre (43,560 square feet) requires 325,850 gallons of water (this is called 1 acre foot of water). One acre inch of water equals 27,154 gallons.

The other method of providing water for your farm is simply to conserve every drop of rain that falls on it. This has the advantage of being sustainable and cheap. Even if you plan to do some irrigation, conserve water to keep your costs to a minimum. Here are some management strategies to save water:

❖ Make sure your soil is worm-friendly (see Earthworms on page 56), which helps hold water in the soil.

❖ Make sure your land has crops (including pasture and cover crops) on it during rainy seasons, to prevent runoff and to hold water.

❖ Plan your field rows to retain water rather than lose it — this means that if you plant on hillsides, for example, always terrace your rows to prevent runoff.

❖ Always provide a place for water flow to accumulate, rather than leaving your farm.

Ponds

Even farms with the best soil will have rainfall that causes runoff. Ponds are an effective method of storing water for later use. Ponds — or tanks, as they are called in the western states — can be used for many purposes: crop irrigation, watering livestock, home water supply, fire safety, erosion prevention, wildlife habitat, swimming, and fishing. A pond should be fenced to water livestock below the dam and keep the water pure and clean for human use.

If stocked with fish, the pond can add income to your farm through fee fishing. You can also consider aquaculture in on-farm ponds. Freshwater shrimp can be raised in ponds as far north as Kentucky, and crayfish even farther north. I raised catfish for many years, with an on-farm fish-processing plant (consisting of a room with two sinks and a freezer). I sold the fish to local grocery stores.

Cover Crops

We briefly discussed growing cover crops in chapter 4, but I would like to focus on the weatherproofing benefits of this practice. Cover crops slow down both wind and water erosion, as the thickly sown plants form many little surface dams to slow water and soil movement. The root systems do the same below the surface. With slower movement, the water has a chance to soak in, allowing plants to have an increased supply of water later in the season. Cover crops also stop nutrient leaking. Winter rye, for instance, sops up excess nitrogen in the soil, while hairy vetch is good at hanging onto phosphorus. The plants release these nutrients when they are plowed or disked under as green manure.

Timing

When you plant your crops is critical in regard to saving moisture. Every time you work the soil for planting or cultivation, you do two things. First, you lose some moisture from the soil. Second, stirring the soil increases the microbial action within it, burning up nutrients and organic matter.

This means that green-manure crops need to be turned under and left to rot for about 2 weeks before you plant corn. The green-manure crop from last fall is a storage base that holds plant nutrients in place over the winter. They become available to your new crop when you turn them under. However, turning them under does cause loss of moisture. It is important to turn under fall crops early in the spring to keep them from using the moisture needed for your new crop year.

Creating a Microclimate

You cannot change the weather, but you can modify its effects by creating microclimates. A beneficial microclimate is an area that is protected from the worst effects of wind, weather, and temperature. It often tends to stay frost-free when surrounding areas are not, or it may offer protection from a hot sun in July.

Examine your farmland to see if you have any natural microclimates. These will be sheltered areas, possibly in a depression where the wind does not blow (although depressions often suffer from frost problems) or on a low hillside that is not as affected by wind or frost. If possible, walk over your property on a frosty morning in late fall or early spring and see if any areas are less affected — or not affected at all. These will be good places to utilize. You may wish to modify them further, to enhance the effect.

A little record keeping goes a long way. Buy a soil thermometer and test soil daily. Common plants such as lilac enter the first leaf and first flower stages of growth at specific temperatures and weather conditions. Checking temperatures and dates, and comparing these over the years, will make your farm more successful.

If there are no natural microclimates, create your own. In planning a microclimate, the best place to start is with a windbreak, which I will discuss in more detail in the following section. The windbreak controls a large microclimate, which sets the stage for a more subtle microclimate in your garden or planting area. Keep the ground protected, for example, with plastic or straw, to create a warmer environment for seedlings, hold extra water, and reduce weeds. (I tend to avoid plastic because it creates a mess when it breaks down; straw, on the other hand, breaks down into extra organic matter.)

Air flows just like water and settles in low spots. A raised bed might be just high enough to keep plants away from frost. Raised beds also provide warmer soil temperatures in the spring, sometimes by as much as 5 to 10 degrees.

It is also possible to create a microclimate for an individual plant by use of season extenders or similar tools. Research by D. R. Paterson and D. R. Earhart at the Texas Agricultural Experiment Station reported that when tomatoes were grown in black plastic with cages and then the bottoms of the cages were wrapped with common roofing paper, losses due to wind and hail decreased by 46 percent. After 5 weeks, there was a 62 percent increase in growth, with 86 percent more marketable fruit and a 49 percent increase in total yield. This data may not apply to all tomato varieties, but the method is worth a try if you have a wind problem early in the season.

Windbreaks and Sustainability

> **PRINCIPLE:** *Plant or create windbreaks to prevent heat and moisture loss.*

Wind can be a problem, as it leaches heat and moisture from crops and livestock. Therefore, crops and livestock protected from wind use moisture and nutrients more efficiently. One way to remove wind as a problem is through a windbreak.

Windbreaks reduce wind velocity, slow wind erosion, and create microclimates. Soil particles move when wind speeds are 13 miles per hour, 1 foot above the ground. When wind flow removes fine soil particles from one field, organic matter and nutrients go with it, so you have to increase fertilizer input or accept lower economic production in your crops.

Decide where to plant your windbreaks based on the direction of the prevailing winds. In most areas of the United States, you'll need to plant windbreaks to the north and west of the area you want to protect. There is more information on height, length, and density of windbreaks in the section that follows. In general, though, windbreaks should be at least 100 feet longer than the area you want to protect, to prevent wind from whipping around the windbreak edges. For home protection, windbreaks should be

100 to 150 feet long, located no more than 300 feet away from the house, yard, and outbuildings.

A wind velocity of 5 miles per hour is slowed to ½ mile per hour in the lee of a windbreak — a 90 percent decrease. A 30-mile-per-hour wind will be reduced by 50 percent, to 15 miles per hour, by a windbreak. The lee can be a distance of up to thirty times the height of the trees (see next section).

University research shows that heat energy savings of up to 40 percent are possible when you use windbreaks. One study showed that a house with a constant temperature of 70 degrees Fahrenheit protected by a windbreak requires 23 percent less fuel than a house exposed to the full sweep of the wind. Permanent windbreaks on 40-acre fields at the University of Nebraska's Mead Research Station increased soybean yields by 18 percent, corn yields by 20 percent, and wheat yields by 22 percent.

Windbreaks are a worthwhile investment. James R. Brandle, Bruce B. Johnson, and Terry Akeson, in a University of Nebraska study, found that "a full windbreak occupying 5 percent of the

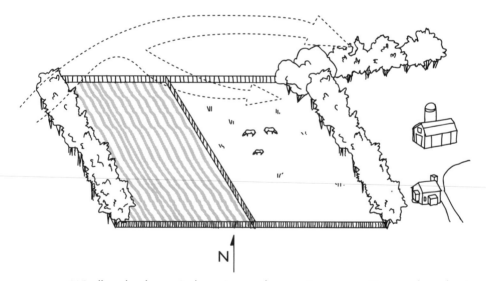

Windbreaks slow wind erosion, reduce water evaporation, and moderate temperatures to protect your crops and livestock.

field is economically viable" and will "more than compensate for the cost of establishing the windbreak and the loss of output from acres taken out of production."

Kinds of Windbreaks

To determine the best kinds of trees for windbreaks, consult your local university extension office or state Department of Agriculture. You will choose trees based on their lifespan, density, growth patterns (for example, evergreens do not shed needles, and thus provide more protection later in the year), and height. Trees or shrubs can be planted in single rows or in mixed groups.

Height. The height of a windbreak determines the protected area. A rough formula predicts wind speed reductions in an area of 2 to 5 times the height of the windbreak on the windward side and up to 30 times the height on the leeward side. This means that with 30-foot trees, the protected zone spans 60 to 150 feet on the side the wind is coming from and up to 900 feet on the side away from the wind.

Length. The length of the windbreak determines the amount of area receiving protection. According to windbreak expert James Brandle, the maximum efficiency for windbreaks requires that the length be 10 times the height.

Density. Varying the density of the windbreak can influence what you do with it. For instance, 25 to 35 percent density is best for even spreading of snow across a wheat field but will not control soil erosion as well as a 40 to 60 percent density. Evergreen trees, among them Eastern Red Cedar, are fairly dense. They are good choices for windbreaks because they don't lose their needles in wintertime, unlike deciduous trees (broadleafs) such as oak.

The main advantages of tree windbreaks are their height (30 to 50 feet) and longevity (50 years). Trees are not the only wind-break materials, however. Perennial grasses and legumes can also be planted in strips. As well, small grains like rye, wheat, and oats can be planted in strips to slow wind erosion, conserve moisture, and provide a suitable microclimate for vegetables that need more wind protection, such as onions and early-planted cole crops. Corn rows planted in a field of soybeans or wheat grass in wheat fields are also beneficial. Two rows of corn are usually alternated with sixteen rows of soybeans or four to sixteen rows of corn and soybeans are planted in alternate strips.

Wheat grass is planted in double rows spaced 36 inches apart, every 48 feet across the wheat field; wheat grass grows about 5 feet tall. Annual plant windbreaks can be placed around an area perpendicular to the prevailing wind direction, just as with trees. In most areas of the United States, this means placing

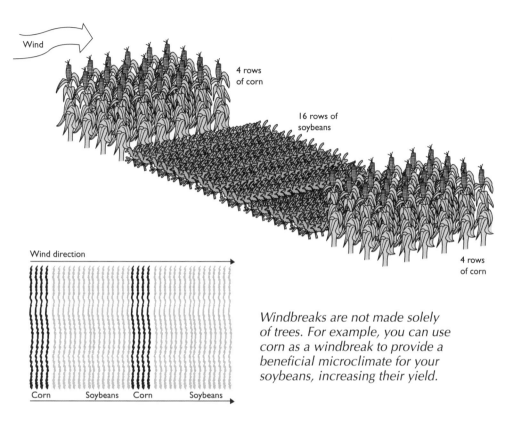

4 rows
of corn

16 rows of
soybeans

4 rows
of corn

Wind

Wind direction

Corn   Soybeans   Corn   Soybeans

*Windbreaks are not made solely
of trees. For example, you can use
corn as a windbreak to provide a
beneficial microclimate for your
soybeans, increasing their yield.*

the windbreak on the north and west sides of the area to be protected. If there is no predominant wind direction, Laurie Hodges
and James R. Brandle recommend planting annual windbreaks
with the rows closer together, following the land contours or in a
serpentine pattern, to slow the winds and protect the other
plants.

Strips of small grains can also be used as lures to keep
insects away from your cash crop. Insects prefer the small grains
and will remain in them, rather than eating other crops. It is sug-
gested that the strips of small grains be 45 to 60 feet apart in
your field and of a width appropriate for your equipment.

Extending the Season

Season extenders are a way to get crops growing early or to lengthen their growing period late in the year, to allow you to have crops when other farmers do not. This is an easy way to reduce competition and increase your profits. Greenhouses, cloches, tunnels, and floating row covers can all be used to extend your growing season.

SOLAR GROWING

According to Shane Smith, author of the *Beautiful Solar Greenhouse*, food production in greenhouses actually had its beginnings in a medical prescription for the Roman emperor Tiberius Caesar. He was told by his doctor to eat a fresh cucumber each day — even in winter — so his workers created a pit in the earth and covered it with a transparent stone. Manure may also have been used for its heat-producing qualities.

Shane Smith is a founder of the Cheyenne Community Greenhouse in Wyoming. Cheyenne is a cold place to live, and it ranks fourth highest in the nation for yearly average wind speed. The Greenhouse, a community project, is 5,000 square feet and is 100 percent passive solar heated. Two hundred 55-gallon black-painted drums containing water store solar heat for the greenhouse. The north, east, west, and roof are heavily insulated. The Greenhouse has never required back-up heating, even during the coldest weather.

Anna Edey's year-round solar greenhouse, Solviva (Swedish for "sun life"), is located in Martha's Vineyard, a small island off Cape Cod, Massachusetts. Solviva houses both crops and livestock, and uses only the sun and animals' body heat for warmth. Anna has 100 chickens laying eggs in the greenhouse to supplement the 80 to 100 pounds of salad greens that are produced every week. She grosses $70,000 to $100,000 per year; the 104-by-28-foot structure cost only $30,000 to build in 1983.

You can grow lots of bedding plants in a small greenhouse — up to 5,000 plants have been grown in a 16-by-16-foot space. These two examples of permanent freestanding greenhouses have production figures of 3 to 4 pounds of food crops per square foot. Your efficiency will improve as you learn, so don't be afraid to start.

I have a closed room in my machine barn in which I start tomatoes, peppers, melons, and herbs early. I set hooks in the ceiling and string fluorescent lights about 4 inches above a series of folding tables. (*Note:* Fluorescent lights must be replaced at least every other year, as their intensity does fade.) I plan planting dates to allow seeds to sprout and grow to about ½ to 1 inch 2 weeks before frost-free dates in spring. Once they are growing well, I move them outside during the day, placing them in a ring

---

If these examples have you interested in greenhouses, but you're not sure about the investment, there are lots of ways to extend your growing season cheaply, and to get a feel for solar growing:

❖ If you have a south-facing house, barn, or shed, just lean a window, a sliding glass door, or a piece of plexiglass against the building. Cover the ends with plastic or straw bales, and you have a miniature greenhouse for early greens or starting plants for transplanting.

❖ Lay out several straw bales in parallel rows. Cover the rows with window panes, or 2-by-2- or 2-by-4-inch wood frames covered with plastic. Placing milk jugs painted black and filled with water between the bales will help dispense heat at night when things cool down. Bales placed crossways at the ends of the two rows of bales will hold in even more heat at night, and provide excellent protection against the wind.

❖ A cold frame is simply a wood frame covered with a window pane. These mini-greenhouses are easily transportable.

❖ Milk jugs make great tiny greenhouses to start plants in the field early in the season. You can start sweet potato slips in April under milk jugs. Just cut off the bottom of the jug and place it over the plant, pushing it far enough into the soil to hold against winds.

❖ Raised-bed gardens with hoops of PVC pipe covered with plastic make good season extenders. Raised beds are large frames of wood (or tire rubber), usually placed with rebar rods driven beside them for structure, then filled with soil. A typical raised bed might measure 5 by 30 feet. The hoops and plastic can also be used on flat-ground gardens.

GROWING POTATOES

Plant potatoes as early as possible in the spring. Hill them when they are 3 or 4 inches tall. (To hill a potato, mound the earth around it until the hill is 8 or 9 inches high and 12 inches wide at the base.) Then spread straw heavily on top, which not only conserves moisture already in the ground, but also soaks up any new rain and virtually eliminates weed problems.

of square hay bales to protect them from the sun. This hardens them off (adjusts them to outdoor temperatures) and prepares them for planting. On the first frost-free date, I transplant them.

Season extenders will also save moisture. Covering garden beds with black plastic or mulch early in spring will hold in moisture enough for you to grow a good tomato crop with little or no added water. Mulch such as straw keeps the ground moist through the hottest summer months.

Caring for Livestock in Winter

PRINCIPLE: *Let your animals learn to cope with winter.*

Your livestock may need better shelter in winter to protect them from the wind. Overall, though, the best thing you can do for your animals is to let them learn to cope. In northern areas, cattle will learn to graze through snow and can use the snow as a water source.

We are apt to do too many things for our animals that are not conducive to developing good, hardy breeding stock. Stock that can survive on the land on its own will minimize our input in time and supplies — allowing our labor to be used elsewhere and saving money.

Providing Water

Sheep obviously need some water, but can usually do quite well on range or pasture with what nature provides. I have seen my own Katahdin Hair sheep go without water for up to a week in the winter, even though it was available to them.

Cattle, hogs, rabbits, poultry, and horses require a source of freestanding water to drink. This means that you must check your water source each morning and break any surface ice so the water can be reached.

If you have only a few head of livestock, the best way to water them in winter is with buckets of warm water. If your livestock numbers are large and spread out, an automatic energy-free waterer saves both time and water. You still need to visit your livestock every day, though, to check that the device is working.

Healthy livestock can easily withstand subzero temperatures if they are relatively dry and protected from the wind. In this example, the hills serve as a windbreak.

Providing Feed

Livestock require more feed in winter, as pasture grasses decline and their nutritive value decreases. Plan to store crops such as ear corn and hay for your winter feeding. You might also investigate cold-hardy forage crops that will allow your stock to graze into late winter — or all winter, in the South.

❖❖❖

FOOD FOR THOUGHT

Sustainable agriculture is an ongoing discovery of how things work and interact on your farm. Saving every drop of water that falls on your land involves using tillage methods like fall-sown cover crops, windbreaks to slow moisture depletion, and terraces and dense windbreaks to distribute water and snow. By seeing the interconnectedness of how one thing affects another, we can put together a series of practices that will make a farm sustainable and successful.

❖❖❖

PLANNING
and
Marketing

CHAPTER SIX

Your Goals and Farm Planning

Why do you need goals for your life and your business? The answer is simple: If you don't know where you are going, how will you know when you get there? The more goals you write down, the better your chances of success.

Goals vary. There are short-term, intermediate, and long-term goals. There are personal goals, family goals, business goals, and farm goals. A family goal might be to develop a system of farming that allows you to spend more time with your children. A personal goal might be to have enough farm income to quit your town job and farm full time. A business goal might be to achieve a 20 percent return on your total investment. A farm goal might be to develop a system of farming that is economically viable, socially acceptable, and environmentally sound.

As you can see, goals can overlap, and different goal aspects can be present in a single goal. The important thing is to figure out what your goals are, so you can plan your farm.

≺ *Diversified crops give you a multitude of markets — and marketing options — throughout the year.*

Planning

A young couple, Earnie and Martha Bohner of Persimmon Hill Berry Farm, spent 3 years accumulating budgets and information on various crops and livestock before they chose the enterprises they thought suited their soils, location, and monetary goals.

After determining what they wanted to grow, they established a 10-year plan, a 5-year plan, a 3-year plan, and a 1-year plan, broken down into what to do every 2 weeks.

In the process of studying various enterprises, they attended conferences and seminars, bought books, and talked to farmers. This intense study of the crops they were interested in exposed them to different tillage methods, irrigation methods, and marketing methods. Some of the knowledge that they gained from farmers, instead of from a book, changed their goals or the order in which they chose to accomplish their goals — some would take longer and some could be achieved more quickly. The Bohners eventually decided to start with U-pick berries and shiitake mushrooms. (More information on their operation can be found in chapter 7.)

Earnie and Martha worked full-time jobs while accumulating their information. Will you need 3 years of study to put together a perfect plan? That will depend on your available time and persistence for the project. You might need only 6 months to a year — but remember, "Nobody plans to fail, but many fail to plan."

PRINCIPLE: *Planning and flexibility are two keys to farm success.*

In addition to planning, you must be willing to be flexible and shrewd. Take advantage of new ideas, new markets, and any opportunities that arise. Right after college I was a foreman on a sod-laying crew, and the bluegrass sod we were cutting belonged to an elderly farmer who used to tell me about the Depression. He said fat hogs were selling for 3 cents per pound. One day there was a really bad ice storm. He loaded his hogs carefully and hauled them 125 miles to the

St. Louis market. Because of the storm, there were too few hogs there to satisfy the buyers that day, so his hogs brought 5 cents a pound. He took advantage of the weather to increase his profits.

Monetary cycles and cash-flow shortages can make any enterprise — no matter how much you love it — into a living hell, draining both your enthusiasm and your pocketbook. A sustainable, diversified farming operation gives you the best chance of earning a living from the farm and still having a life. Yes, I said the best chance — no guarantees. The best and most frugal farmer in the world cannot survive unless he or she can pay the bills, good weather or bad. Planning and flexibility will allow for a steady cash flow.

A Planning Example

Let's take some space on your farm for a planning example, without considering budgets or marketing. You have decided to plant some vegetables.

Slope. First, consider the land itself. Most vegetables need a warm environment in order to prosper. A gentle slope is a good place to plant. Why? Well, the sun is the primary source of heat for soil. The heat is greater in the summer when the sun's rays hit the soil directly; in wintertime, the sun's rays hit the soil at a slant, because the sun is lower in the sky as a result of the tilt of the earth. But on a slope, the sun is able to make direct contact with the soil for a longer period of time due to the angle of the land. This means it will warm up and dry out sooner in the spring, and thus can be planted earlier.

In general, in the northern hemisphere, trees grow on northern slopes and grasses on southern slopes. Southern slopes tend to be warmer and drier than northern slopes.

Moisture. The angle of slope of a piece of property can modify the climatic conditions on that land. On steep slopes, more rain runs off instead of penetrating the soil. Thus, the soil is drier and provides less water for your crops. On sloping ground, there is more moisture toward the bottom of the slope. Your choice of vegetables or field crops should take advantage of this moisture.

Soil. You will also need to consider the soil of the slope. Sandy soils are porous and warm up quickly in spring. Clay and clay loam soils hold water and dry slowly in spring. Evaporation of water from the soil also reduces the direct effect of the sun's rays.

Planting. Now consider how you will plant your early-vegetable crop. If this slope is more than 2 percent, you want to run your vegetable rows perpendicular to and around the slope, each row of plants forming a dam to catch water and slow the erosion of soil. If the slope is greater than 5 percent, different growing methods must be applied, such as strip-cropping (see box).

Management. You'll also need to consider the effects of your other farm operations. For example, running livestock on the slope the year before you plant, or feeding hay to stock and concentrating the manure on the slope, may furnish all the fertilizer you need — virtually free of charge.

This may seem like a lot of decisions to make just to plant some vegetables, but things will get easier. Once it becomes a habit for you to consider all your options, some decisions become almost instinctive.

STRIP-CROPPING

Strip-cropping is used on land that is moderately sloping. Crops are planted in alternating strips, each strip being from 4 to 12 rows wide, horizontally across the slope. The idea of alternating strips is to allow for a thicker crop (hay, for example) to prevent water flow from eroding the slope, while still utilizing the slope for more high-value crops.

Vegetables, for instance, are surrounded by bare ground or mulch, so they need a thick crop alternating with them. The alternating strips might be corn and soybeans on a gentler slope, corn and hay on a steeper slope. Crops are rotated in the strips: for example, 2 years of corn, then the corn strip is sown to a hay crop, then the hay strip is plowed under, furnishing organic matter and soil fertility for the next corn crop. The hay strips may be grazed, using electric fence to keep sheep or cattle out of the corn. (See chapter 4 for more on crop rotations.)

Setting Goals

Following is a list of sample goals. If these goals do not seem to fit your plan, make your own. For instance, the goals for a fee-hunting ranch would certainly different from those I have presented here. Your own list may be much longer or shorter, but the more effort you put into it, the greater your chances of success.

Long-Term Goals

Long-term goals, 10 years or more, should be written down and reviewed monthly to see whether you are proceeding in the right direction or if you have veered off track. As time goes on and your experiences or markets alter your plans, you may change your goals. Be sure to write down your altered goals and spend some time planning how accomplishing the new goal will differ from what you had previously planned.

Write down broad-scope goals as long-term goals. Some examples:

Goal 1. **We want to farm.** This is a good broad goal, but if it is your starting point, you have a lot of research to do.

Goal 2. **We want a sustainable, diversified, profitable farm because it gives us the best chance for success.** All of these are necessary for success, but the research still needs to be done.

Goal 3. **We want a pasture-based farm; or we want a diversified crop and livestock farm; or we want to raise purebred animals.** Your research and needs will help you choose a direction before you decide on specific crops or livestock. The process of establishing a successful, respected farm takes several years.

Goal 4. **We are from the city, so I will work part time for a farmer, maybe on weekends, to learn basic skills and ask questions, questions, questions, so that I can gather an idea of what I want to do. I will work for free if necessary to achieve this goal.** Everyone should have some method(s) of gaining knowledge as a long-term goal. You should be learning about farming from your farm for the rest of your life. Sustainable agriculture is an ongoing process of discovery.

Goal 5. **We — the family unit, wife and children, my brother and I, my neighbor and I — will be in total agreement about our monetary and labor goals.** It is great to say, "Let's raise blueberries." But who does the planting, the pruning, the marketing? Will these chores be split? How? If the unit cannot agree at the outset, the project is doomed to failure. This is a long-term goal because it is broad in scope. Saying "Sheila will plant the blueberries this spring" is a short-term goal, but planning the family's overall direction for an operation is long term.

Goal 6. **We will reduce input costs. We want a type of farming where crops and livestock complement one another, and return a greater profit to us while improving our soil, our net worth, and giving us a higher quality of life.** Reducing input costs probably involves either improving soil fertility (healthier soil requires fewer inputs) or increasing diversity. Soil fertility is a long-term goal (see Goal 16). Using diversity — a combination of crops and livestock that complement each other — can, with careful planning, reduce costs. For example, hogs eating your own homegrown corn reduces feed expenditures. Improving soil and net worth and improving quality of life should be separate goals. As individual goals, some of these might be achievable in a shorter time frame.

Goal 7. **We want to sell products that fit the organic market niche.** This is a more specific goal that may be achievable in a shorter time — but until you know what you are doing, treat it as a long-term goal. What does it take to be certified organic? Will crop yields be lower, the same, or higher? Remember that being organic does not make your farm sustainable, and vice versa.

Goal 8. **We will try any new project on a limited basis and keep good records. I will not bet the farm on hearsay or hype from universities, newspapers, magazines, and so on. I will first test it on my farm and also talk to other farmers who have done it.** Once again, the breadth of scope makes this a long-term goal. Any individual project to which you apply this principle will be a short- or intermediate-term goal. If the project works, it may revert again to a long-term goal. Often, it is worthwhile to write down what may seem like plain common sense as a goal. This allows you to come back to it and think about it periodically.

Goal 9. **We will buy only that machinery that we absolutely need.** Machinery depreciates from the day you buy it, and when it gets rusty it really goes down in value. However, if you have good tool skills, used machinery and tools can be a really good buy.

Goal 10. **We will keep good records and know our cost of production per bushel, per plant, per animal, per acre.** Records are a day-to-day short-term goal, an intermediate goal, and a long-term goal. To be really valuable, records must be kept on a long-term basis and evaluated periodically. Eventually, they will allow you to make profitable long-term decisions, based on the production capability of your farm. If you do not know your cost of production and you market directly to the consumer, how will you know what to charge?

Goal 11. **We will use 5-year averages for budgets with a cash flow of 120 percent.** Say you are selling fresh brown eggs to neighbors, market customers, and a local restaurant. If your expenses are averaging $100 per week (you will average out your expenses over a 5-year period, adjusting for any unforeseen long-term changes, such as switching from grain-fed to range-fed), then according to your goal, your projected revenues should be $120 per week (120 percent of $100). The extra $20 is your profit, to apply to living expenses, savings, farm upgrades, and so on.

Goal 12. **We will reduce or eliminate as many risks as possible in our operation.** Agriculture has major problems or risks that affect profitability, such as weather, prices, government regulations, disease (plant and animal), weeds, and poor soil fertility. If 20 percent of your income derives from overcoming these obstacles and

80 percent comes from your value-added direct marketing efforts, you have reduced your risk in farming. Value adding is important.

Goal 13. **We will continue to educate ourselves every year.** Make an effort to attend and participate in conferences and seminars every year. Read as much as possible on research for your particular crops and livestock. Use this new information to update, change, or eliminate some goals after test trials on your farm.

Goal 14. **We will direct-market as much farm production as possible.** Selling retail brings higher profits and more customer contact. Selling direct also gives you the chance to sell other products you raise to customers.

Goal 15. **We will have a marketing plan with sales every month of the year.** Sales are essential every month, because you have bills every month. This is a short-term, intermediate, and long-term goal. Even if it is ultimately achieved, periodic reevaluation is required.

Goal 16. **I will ceaselessly strive to build soil fertility by using cover crops, green manures, livestock, strip-cropping, and other soil-conserving tillage methods.** This is a lifetime goal. Improving soil fertility is a long, slow process, because every crop you raise on the land uses soil nutrients. You must carefully balance your choices of crops, livestock, and management techniques to improve soil fertility. You will probably need a lot of trial and error to learn how your soil reacts over the years.

Medium or Intermediate-Term Goals

These goals are what you will seek to accomplish in 3 to 5 years. They must be specific tasks, and will often be broken down from long-term goals. Some medium goals:

Goal 1. **I will purchase breeding stock, machinery, buildings, and so on as low-interest loans with 3- to 5-year terms.** Operating loans and lines of credit are for items to be used up and paid for in 1 year, such as seed, fertilizer, and veterinary supplies. A good ratio for the bank is 2:1 — in other words, $2 worth of assets for every $1 of liabilities (debt). Of course, it is always better to avoid any debt (see Goal 3).

Goal 2. **I want to plant blueberries and sell them at the farmers' market and to local stores.** Keep in mind that it will take at least 3 years for any production and 5 to 6 years for full production. It would be wise to consider single-season vegetable and/or field crops to give you income while your blueberries are maturing. This will also allow you to develop some loyal customers at farmers' markets and stores in advance, so you will have a developed market for your berries.

Always confirm market interest for a crop before you plant. Because there are no immediate returns on your planting, these types of crops will need 3- to 5-year loans, not operating loans, unless you can get a revolving line of credit. Revolving credit pays up and pays down, usually for a specific amount, say $10,000 to $40,000.

Goal 3. **I will avoid short-term debt like the plague.** With a good ratio of 2:1, you can weather some tough financial storms. If you do whatever it takes to make the payment on your long-term loan at the ratio of 2:1, you usually have about 50 percent equity in your operation. When short-term debt piles up, it gets harder and harder to get out from under. This is also a long-term goal, because you will always want to avoid debt.

Goal 4. **As I add new enterprises to my goal planning in the next couple of years, I will compare "apples" to "apples" by taking a look at the gross income per acre minus the direct cash costs per acre, which equals the gross margin on profit per acre or per head of livestock.** By leaving out fixed costs and mortgage payments, for example, you analyze the direct profitability of the enterprise on its own merits — or lack thereof. Looking at income and expenses for 3 to 5 years will give you a good idea of what you can do and how much it will cost.

Goal 5. **I will set in operation some risk management for my farm.** What kind of risks will you have in, say, a U-pick operation or a breeding-stock operation? What happens if you get sick and cannot work? Do you have a reserve in case of drought or flood? Can you buy crop or livestock insurance?

Use life and health insurance to your advantage. In 1999, the government began gross revenue insurance, which covers almost all crops and livestock. If things go awry, you are guaranteed a certain gross revenue per acre or a certain yield (bushels per acre), all for a 20 percent higher premium.

You can control costs and increase output for increased gross income, but you have no control over the government and politicians and the laws they might pass that will affect your bottom line. Make a plan to deal with unexpected changes.

Goal 6. **We want to convert our cattle operation to a management-intensive grazing operation.** You've estimated that it will require 2 years to build fence and lay water lines. How will this affect the cash flow? Will it generate more income the first year or second?

Short-Term Goals

You will seek to achieve short-term goals within 1 to 2 years' time:

Goal 1. **We have existing portable hog houses and will purchase feeder pigs to sell as whole-hog sausage in the fall.** Any short-term goal will probably spawn other short-term goals. In this case, for instance, decisions need to be made on numbers and types of pigs, feed, and marketing. Other goals prompted by this one might include "We will buy 4 crossbred (Yorkshire x Hampshire) feeder pigs from Bob Smith in late August," or "We will mix the sausage at 80 percent lean and 20 percent fat, and sell at a retail price of $2.50 per pound (120 percent of expenses, from the long-term goals)."

Goal 2. **We will build two portable chicken tractors and raise about 200 birds for sale as meat in about 8 weeks.** Chicken tractors are portable cages that allow chickens to graze in your garden between the rows and fertilize it, without eating your plants (see page 83).

Goal 3. **We will buy ready-to-lay pullets and start an egg-laying operation.** Long-term and intermediate goals are often loosely defined or even philosophical, but short-term goals require that specific decisions be made immediately. How many pullets? What feed? What production level will be managed for? (Seventy to 80 percent is a good starting level.) What will we charge for a dozen eggs?

Goal 4. **We will figure in time for rest and recreation this fall.** Vacations, entertainment, and rest are necessary to renew your body and spirit; they also allow some family time away from the concerns of the farm. Make sure that rest and recreation are always on your list of goals.

Achieving Your Goals

Setting your long-term, intermediate, and short-term goals is like using a map, and the goals are your destination, depending on where you are at on the road. You must have a plan. Some people say, "I have a plan in my head" or "Planning is boring. I want to raise ostrich now!" These people will fail without an excessive amount of luck. Planning is the easiest and least risky way to make your small farm successful and profitable.

I cannot count the number of times we at *Small Farm Today* have received calls from farmers who say something like, "Well, I just finished growing my specialty popcorn. How do I market it?" We try to suggest possible marketing avenues, but if they have not been planned in advance, the farmer's family will likely be eating a lot of popcorn, or whatever else they raised without planning on how to dispose of it.

Successful farmers write down their goals. They will know how to raise their product, what their cost of production is, and, with careful planning, where they are going and how to get there.

Developing Your Farm Plan

Your farm plan should be like a small business plan. Consult with a local Small Business Development office for information on forming a business plan. You will need this both for loans and for your own planning purposes.

A business plan explains how you will achieve revenues from sales and how you will spend expense money. In particular, it demonstrates how you will repay loans and interest. It also shows how your business fits into the marketplace, examines who your customers are, how you will price your product, and what future developments may occur. A good plan accounts for variables that may affect income. Farms have some variables to consider that are not typical of other businesses:

Weather. Weather will always be your first and biggest problem, but with cover crops, buildup of organic matter, and some limited drip irrigation or other limited water usage, you have eliminated all but the temperature aspects and catastrophic problems like floods and tornadoes. Although you may not have specifics on weather in your plan, you need to consider its effects and devise ways to counter them (add-on value, for example).

Weed, disease, and insect problems. These are the target of the production strategy of your business plan. You should employ a production method that lowers purchased inputs, such as chemical pesticides. Weed, disease, and insect problems can best be handled by building the organic matter and fertility of your soil. Building soil fertility must be one of your broad, long-term goals. The healthier your soil, the fewer problems you will have with weeds, disease, and insects. Any problem that can be cured with a "spray-to-death" program can be cured in another way that is less expensive and less harmful to the farmer and the environment. And if you plan to be organic, another cure *must* be found.

Management. Japanese farmer Masanobu Fukuoka calls his technique the do-nothing method. What he really means is to manage your farm in a manner that reduces costs, making it sustainable because it becomes profitable. Jim Gerrish, grassman, and Allan Nation, editor of the *Stockman Grass Farmer* newspaper, changed the term "intensive grazing" to MIG — management-intensive grazing. They were underscoring the importance of management and the thinking process in farm success.

Fancy equipment or high-priced show livestock does not necessarily make your farm profitable. On my own farm of 80 acres, I have one old tractor, a tiller, and a few implements. I grow about 5 acres of open-pollinated corn as feed for my sausage hogs (bought cheap as feeder pigs) and poultry. This is primarily because these animals are not as good at converting pasture to profits as are my Katahdin Hair sheep, which I have more of and which use more grass — an annual renewal source. These are not show sheep, but sheep bred for my farm conditions. Pasturing poultry and hogs can certainly reduce your feed costs and give you a better, healthier product to sell, but poultry and hogs are simple, one-stomach animals. Sheep and cattle, in contrast, are ruminants (they have four stomachs), with the ability to consume larger amounts of roughage. They can be raised to a marketable weight on forage alone. Low inputs, efficient livestock, and direct sales keep my operation profitable.

Government regulations. Like it or not, you must be involved in meetings about legislation. The general population is four to five generations removed from the farm and this includes our legislators, yet they pass laws about an industry with which they are unfamiliar. It is very important that legislation that affects direct marketing be addressed so the small farmer has a chance to compete with the big boys.

Farmers are less than 2 percent of the total population. We feed the other 98 percent. Every farmer must be a spokesperson for our industry. You cannot live in the woods doing your own thing, so to speak, because you can be legislated out of business if you are not aware of what is going on.

Adding value. Value-added products include jams, cuts of meat, sausage, rubs, shampoos, craft projects, and much more.

Masanobu Fukuoka is a Japanese farmer famous for his natural farming methods, which he described in his books *One Straw Revolution* and *Natural Farming.* He grows rice and winter grain on a ¼-acre field, and has harvested 22 bushels each of grain and rice from that field — without any inputs at all. Fukuoka's philosophy translates into a thinking-type agriculture. American agriculture is steeped in tradition, but not in thinking about *how* and *why* we do the things we do.

Fukuoka, born in 1914, categorizes farming into three basic types:

Mahayana **natural farming** is perfection — the farming of Nature. Natural farming follows the Buddhist philosophy of *mu,* or nothingness, by obeying a "do-nothing nature." Fukuoka says Nature has everything; no matter how man struggles, he will never be more than a small, imperfect part of its totality. This notion of farming is more of a philosophical ideal than a farming method attainable by man.

Hinayana **natural farming** tries to remove human knowledge and action and utilizes the pure forces of Nature. It is the natural farming method that results when man works with Nature; this is the method that Fukuoka uses. This farming method takes advantage of natural occurrences. It is a "do-nothing" philosophy still, but with allowances for the necessary intrusions of man, such as planting. A typical application of *hinayana* in the United States is overseeding red clover in winter wheat. The clover sprouts in the spring, but its growth is held back by the shade of the faster-growing wheat. When the wheat is harvested in July, the clover is exposed to the sun and grows rapidly, providing a cover crop. Nature determines the interaction of the crops; man assists.

Fukuoka's natural farming has five major principles: no tillage, no fertilizer, no pesticides, no weeding, and no pruning.

Scientific farming uses human knowledge and action in an effort to establish a superior way of farming. Fukuoka believes scientific farming is limited to short-term goals; its achievements may be better in some ways (e.g., yields) but are far inferior in all other ways, and will lead eventually to failure. In comparison, natural farming is total and comprehensive. Fukuoka describes American organic farming as just another type of scientific farming. He says scientific farming is judged on a scientific basis; natural farming should be judged on philosophical grounds.

Not all of Fukuoka's ideas are applicable to American farming, but he is always thought-provoking.

Adding value reduces risk by turning a basic product into a product with additional profit and a longer shelf life. Weather, price, government regulations, and weed, disease, and insect problems are the main obstacles to small farm profits. If the majority of your income comes from a value-added product, you have eliminated most of the farming risk.

Value-added products allow for a drastic reduction in volume, with little change in gross revenues. Most people with value-added products have excess field production (more crops than can be turned into product), which can be sacrificed without loss to the product. The add-on price also protects the gross revenues. It only takes 5 bushels of corn sold as cornmeal to equal the gross revenues of 100 bushels of corn sold at the elevator. (See chapter 7 for more on add-on value.)

Marketing. When success and profit are your broad, long-term goals, direct marketing becomes very important. I will discuss marketing and adding value in the next chapter.

❖❖❖

FOOD FOR THOUGHT

It is my hope that as you read this book, you will commit the basic truths of each chapter to memory to be used every time you make a farm decision or define or redefine your goals. The most fundamental of these practices are to clarify and prioritize your goals, keep them uppermost in your mind, and plan your enterprise with them in mind.

❖❖❖

CHAPTER   SEVEN

Marketing

What are some differences between the marketing strategies used in traditional agriculture and those used in alternative agriculture?

Traditional agriculture grows the crop or livestock and then looks for someone to buy it. Alternative agriculture looks for the market first and then grows what the market wants. By combining this with added value and direct marketing at retail prices, farmers move upward in the marketing process, for much greater profits.

Traditional agriculture depends on "wholesale" prices at elevators and sale barns. However, a farm that sells surplus products above what the family eats is a small business, and small businesses succeed by practicing retail rather than wholesale marketing. Direct marketing is the profit equalizer for family farms.

Traditional agriculture simply sells the commodity that is grown. Alternative agriculture will seek to add value to the product by turning it into a retail consumer item such as jam, sausage, or a wool sweater. By adding value-added products to direct marketing, farmers gain additional profit. It is important to remember that in direct marketing, you are marketing yourself as well as your product.

PRINCIPLE: *A farmer must be skilled at buying and marketing.*

< *Marketing begins by reaching the consumer. Each fall this roadside display of pumpkins attracts the eye of tourists, photographers, and pumpkin buyers as they travel U.S. Route 7 in southwestern Vermont.*

Eight Steps to Identifying the Market

How do we determine what the market or consumers want?

Step 1. Obtain maps of your state and county.

Step 2. Locate your farm on the maps and draw circles with your farm at the center with a radius of 25, 50, and 100 miles.

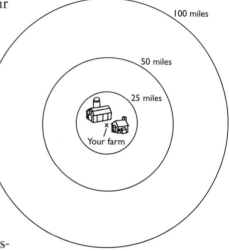

Step 3. Count how many towns, cities, or population centers are within those ranges. Most of your farm's customers will be in the 25- to 50-mile range.

With your farm at the center, draw circles at a radius of 25, 50, and 100 miles.

Step 4. Using your maps, add up the number of people present in your area. Your local agriculture statistics office (check with your local extension office or State Department of Agriculture) may help you compile these numbers.

Step 5. Find out how many alternative agriculture farmers are in the circled areas of the maps, what they are growing, and what they are doing with their produce. If there are already ten strawberry farmers with 10 acres each, the market is probably already saturated for U-pick strawberries — but it may not be if you are willing to process the berries into jams, jellies, or pies. Remember, too, that there will never be an oversupply of the "best" — successful marketing is selling quality. You may have to do some driving and checking of local newspapers, farmers' markets, and your local Extension and agriculture statistics office to get this information.

Step 6. Go to your local grocery store or visit several of them in the population centers. What vegetables and meat products are already available in these stores? Can the stores purchase these items locally or do they import them?

Step 7. After gathering the information, determine what crops and livestock are missing that might have sales potential. Check which products are imported from other states and ask yourself why. Hardly any states are self-sufficient in food production; most grow around 30 percent of what they eat. Growing produce locally helps improve the regional rural economy and is desired by consumers. The market is there for the taking.

> **PRINCIPLE:** *Plan to have produce available when others do not, or offer unique products.*

TRUE STORY

Dwight James, a banker-turned-farmer from Alabama, discovered that his state was a pumpkin-deficient one. In other words, the pumpkins being purchased in Alabama (population: 4.3 million) were coming from Tennessee and New Mexico. The market was there, but Alabama farmers were not taking advantage of it. By competing with imported pumpkins, Dwight could save transportation costs and damages, and rely on "homegrown" civic pride to increase sales. Since this discovery, Dwight is now the president of a sixty-member pumpkin association. Their promotions in the state of Alabama have resulted in the marketing of many varieties of pumpkins for different uses. Because many of these people had never grown pumpkins before, it is clear they are risk takers and price makers — not price takers — which is what agripreneurship is all about.

Step 8. Find out who your customers are. If you have any ethnic groups in your area, investigate growing specialty produce and marketing methods that appeal to them. For instance, certain Vietnamese ethnic groups prefer live black chickens. There are certainly a wide variety of oriental vegetables to grow — or Russian, Amerind, French, Arabic, Mexican . . . . Check farmers' markets and local ethnic stores or associations to talk to people about what they are interested in. Get specific ideas of how they want the product to appear. A satisfied customer is a return customer.

Niche Marketing

PRINCIPLE: *Raise crops and livestock for niche markets.*

Small farmers should raise crops and livestock only for niche markets. This means relying on alternative crops and livestock, or alternative marketing, rather than the traditional types the universities promote. While universities will claim their research is size-neutral, in the real world, that simply is not true. For example, if your university or seed dealer comes up with a new variety of corn that yields an additional 10 bushels to the acre, who benefits the most? The farmer with 50 acres of corn or the farmer with 2,000 acres of corn? The man growing 50 acres of corn gets a 500-bushel increase in his crop. The man with 2,000 acres gets a 20,000-bushel increase, which enables him to purchase new technology at a faster rate than can the 50-acre farmer. The 2,000-acre farmer also has a lower per-unit cost of production and more acres to spread his equipment over.

You might well say, "I have only five acres. I cannot compete, why do it?" Well, you *can* compete, but you need to raise different crops, or the same crops for different reasons. Here's an example.

Selling corn will ordinarily bring you about $2 per bushel. A bushel of shelled corn is 56 pounds, so this amounts to about 3.6 cents per pound. On the other hand, say you take the same corn and add value to it by processing it into cornmeal. You will make $82.88

Curtis Bennett, in Mexico, Missouri, had no land and no machinery. He had been working for other farmers, but wanted an operation of his own. He wished to continue growing the crops he was familiar with, but to make more money and improve his quality of life. He met a farmer with land to lend at the 1994 National Small Farm Trade Show and Conference. Once he had the land, Curtis decided the organic market had excellent potential, and started growing organic corn and tofu soybeans.

He spends about $18 per acre for organic fish emulsion foliar fertilizer for his corn and $12 per acre on his soybeans. He plants the corn following the soybeans so he does not have to buy nitrogen for the corn. He sows about 25 pounds per acre of hairy vetch between August 15 and September 15, which provides an additional 100 pounds of nitrogen for the following year. Sowing in annual rye also helps as a green-manure crop and as a method of weed control.

Yields on the Bennett farm have varied from 33 to 50 bushels for soybeans and been about 115 bushels for corn. These are very respectable yields for low inputs. Curtis's methods have resulted in soybeans that tested at 44 percent protein, a very high-quality bean. The organic industry requires a minimum of 38 percent protein, and pays a premium price for higher-quality beans. Quality tofu soybeans sell for about four times the price of regular soybeans.

Like many farmers, Curtis is always busy. He is a member of the Heartland Organic Marketing Cooperative in Iowa, and has served on its board of directors. The cooperative direct-markets its soybeans, bypassing distributors. About 90 percent of Curtis's beans are sold domestically, and about 10 percent go overseas to Japanese buyers. Most of his corn goes to California for taco chips, burrito wraps, and other natural foods.

Curtis started with nothing but an idea and his own motivation, and now is well on his way to success. Finding the right alternative market and growing the product for it provided him with the key to increased profit.

per bushel ($1.48 per pound) in gross profit. You will have to cover the cost of bags to put the cornmeal in — about 15 cents each for a plastic-lined cloth bag. This means 56 one-pound bags would cost $8.40. This $8.40 will have to be subtracted from the amount you make from selling the cornmeal. Subtracting the $8.40 for bags and the $2.00 opportunity cost (*opportunity cost* is the price of the original opportunity) by selling corn by the bushel leaves a profit of $72.48, which will pay your labor, management, and marketing costs. A good goal is to have 50 percent of your income as net profit — in this case, about $36.25 per bushel.

If each acre yields 100 bushels, selling corn at $2 per bushel will net you $50. Selling it as cornmeal will produce a net income of about $3,625 per acre — including payment for labor and management. By selling the corn as cornmeal — or growing high-dollar, high-yield varieties, or selling "Indian corn" for decorative purposes, or any other number of alternative marketing methods — you can vastly increase your level of profit. Of course, selling 5,600 bags of cornmeal will require much more marketing effort than would hauling the corn to the nearest elevator and unloading it.

Farming Methods and Marketing

How you choose to raise your crops and livestock can influence your marketing choices. Obviously, planting standard commodity crops on a large scale does not lend itself to direct marketing. Even variations of commodity crops (popcorn, white corn, tofu soybeans, for example) usually require contracting with someone to buy the product, rather than direct marketing. This is also true of large-scale operations involving vegetables, medicinal herbs, and industrial crops. Many farming methods, however, lead directly to niche markets. For example, choosing to raise purebred livestock, show-ring livestock, ethnic produce, or your own strains of open-pollinated plants creates a market niche for you. Some varieties of alternative livestock have built-in markets, such as velvet from elk or deer.

If you raise your products by organic or natural means, you will have an excellent marketing approach for consumers. Many people nowadays want to know that their food was raised outdoors in humane conditions, without herbicides, pesticides, antibiotics, growth hormones, or other chemicals. So consider how you will raise your crops and livestock when you are choosing your niche markets.

All About Niche Markets

So exactly what is a niche market? Niche markets are of two types. Sometimes a niche market means a different type of crop or livestock than the two traditional triumvirates of corn, wheat, and beans and cattle, hogs, and sheep. Cut flowers, for example, are a different type of crop that can net as much as $15,000 per acre. Or it can mean produce is marketed differently from the way traditional crops and livestock are, such as at farmers' markets or at a U-pick farm.

Here are some other characteristics of niche markets.

1. Small size. A niche market is small and may be easily saturated. Careful research is required before you choose what to raise. You must assess both the current and potential demand for your product. Because of limited demand, you may choose to raise small quantities of several different crops.

FARMING BY THE SQUARE FOOT

If you are a traditional farmer with a large acreage, you need to set aside a small area to practice intensive growing for direct marketing. Forget about bushels, acres, or thousands of livestock. Instead, think of your farm in terms of square feet. There are 43,560 square feet in 1 acre. If you get 20 cents per pound for the product and produce 1 pound per square foot, the result is $8,712 per acre. The potential of netting $1 per square foot or $43,560 net profit from 1 acre is very real (remember, you are selling direct). The catch is that 1 pound of produce per square foot equals 21.7 tons of produce per acre; you'll need 400 to 1,000 customers to consume that much produce.

2. Add-on value. Niche markets are high-yield, high-dollar, high add-on-value types of crops and livestock. You add on value to a raw commodity by turning it into a different product for a different market, such as selling painted gourds, or marketing freezer beef or whole hog sausage instead of delivering livestock to the sale barn. Niche markets frequently lend themselves to processing-manufacturing, which supplies jobs for the rural community as well as for the farmers. For example, many farmers who started U-pick berry operations find themselves making jams and jellies out of surplus berries or berries not good enough for U-pick sales.

TRUE STORY

> Earnie Bohner's U-pick blueberry operation in Lampe, Missouri, yields about 10,000 pounds per acre and he sells them for $1.35 per pound, or $13,500 per acre. Earnie and his wife, Martha, joined forces with a chef and his wife to develop new products, such as blueberry barbecue sauce, shiitake mushroom sauce, dried shiitakes, and blazons. (Blazons are dried blueberries that sell for $10 per pound — which considerably increases the gross profit on the same per-acre yield.) By putting his berries into small jars as jam, featuring his Persimmon Hill Berry Farm label, for $3.50 per jar, the same 10,000 pounds per acre yields $98,000 in gross revenues.

3. Lack of mechanization. Niche crops are not easily mechanized. The "Suits," otherwise known as speculators and Corporate America, need absolute control of their crops, with mechanization potential for planting on large areas. Consistent production is hard to maintain without mechanization, and unmechanized crops require large numbers of hired help for large scales of production. If a crop takes lots of stoop labor, Corporate America will be less likely to invest in it. A warning: If you see corporations become interested in a particular crop or livestock, get out of it and take your lumps before you are wiped out in the market crash. On a positive note, if you market your production in a way that Corporate America cannot, such as by processing or manufacturing a homemade recipe, you can prosper even with corporate competition.

4. Limited interest. Niche markets are markets often overlooked by mainstream and Corporate America. They may be too small, or require too much labor or management, or they may have been abandoned due to market crashes or lack of profit. You can succeed by finding a different way to market a product that others believe to have limited potential. It is also possible to enter a market during a "crash" period and acquire your stock at a low price. For example, hogs were selling for as low as 18 cents per pound in 1998. Anyone wanting to acquire breeding stock for a hog sausage niche market operation could have bought several hogs, started a breeding operation, and turned some of them into sausage at a selling price of $2.50 per pound (which equals a gross profit of about $2.00 per pound).

5. Controlled volume. A niche market can start with small sales and grow into the volume that you desire. In contrast, corporations like high-volume marketing of products, which yields a lower return but is justified by the volume of sales.

6. Low capital risk. A niche market can usually be gotten into and out of without a large amount of capital expenditures at risk. Niche markets can be long lasting or short lived, which is why you need to be able to get in and out of production. As an example, let's say your niche market is raising purebred hogs on pasture with low-cost farrowing huts to sell to other breeders. If the bottom were to drop out of the hog market, it would be fairly easy to cull your herd to a minimum and convert the rest of the pasture to crops or some other livestock. If you raise confinement hogs in expensive facilities, you are stuck with raising hogs whether or not you profit. These expensive facilities are practically impossible to convert to anything else.

7. Diversification. Niche markets, because of their small size and short duration, almost require that you be diversified. When you sell one animal or one plant, you are not diversified and

Chuck DeCourley, of Columbia, Missouri, was promotions director for *Small Farm Today* magazine. He wanted to start farming but didn't want to spend tremendous amounts of money. After surveying the market, he found that ratite (ostrich, emu, and rhea) prices had dropped following the speculator boom. According to newspaper articles, many farmers, particularly in Texas and Oklahoma, had given up when the "breeder's market" collapsed. Some farmers released their stock into the wild, some sold it at dirt-cheap prices, and a few just killed all their birds. But, as they say, one man's misfortune is another man's fortune.

Chuck purchased emus at low prices and started breeding them. Once he and a friend had enough birds, they started culling them for slaughter. They have marketed emu meat sticks ($1.39), emu sausage ($6.00 for 1 pound), and seven kinds of emu oil products, including a pain rub ($7.00 for 4 ounces). They also sell carved eggs for $150 and up. Chuck is successfully making money off livestock that are being turned loose in some states just to get rid of them. He has since left his magazine career to pursue farming full time.

therefore not sustainable. If you raise small amounts of different produce, it will be easier to sell. For instance, selling 25 pounds of artichokes to a customer is difficult. But you might be able to sell that customer 3 pounds of artichokes, 10 pounds of tomatoes, a 6-pound roast from your Dexter cattle, two dozen eggs, and 5 pounds of apples.

8. Direct contact with consumers. To be a niche product marketer, you must become a salesperson. If you do not believe in your product 100 percent, your customers won't believe in it. If you do not want to deal with the public, you might be better off joining a cooperative marketing effort with other farmers.

Add-on Value

Add-on value is a simple concept. It means that you have processed a raw commodity to create a new product.

Example 1. Berries can be sold as jams and jellies. The jams and jellies add value because they sell at a higher price per unit, but from the same acreage and same yield. Because they can be stored and sold throughout the year, they will produce a larger gross income over a longer period than will fresh berries, which can only be sold in-season.

Example 2. When you sell hogs to the packers, you are selling hogs as a raw commodity. If you turn them into sausage, you are selling a finished product — an add-on value — direct to the consumer.

Fresh berries are perishable, but jams and jellies can be sold throughout the year.

Example 3. Elk are a good add-on value animal. Their antlers are harvested yearly for velvet or craft projects, their meat is sold at a premium, and their hides are turned into handbags, boots, and wallets.

Example 4. Ratites are used for meat, emu oil products, carved eggs, "leather" (from the legs and hides), and feather craft projects.

MORE ADD-ON VALUE IDEAS

❖ **Fruits.** Jams, jellies, barbecue sauces, vinegars, dried fruits, chocolate-dipped fruits, wines, pies, juices, ice cream, crushed ice drinks.

❖ **Meats.** Cuts of meat, sausage, hams, meat sticks, prepared meals (e.g,. chicken cordon bleu, enchiladas, turkey, and stuffing).

❖ **Grains.** Flour, breads, doughnuts, recipe kits, soy nuts.

❖ **Fiber.** Yarns, clothing, felted products, small figures, stuffed animals.

❖ **Decorative crafts.** Christmas or autumn wreaths or swags, dried flower arrangements, cornhusk dolls, painted gourds, gourd birdhouses, painted or carved eggs, scarecrows, feather arrangements, potpourri, herbal gift baskets, pressed flower pictures.

❖ **Vegetables/mushrooms/herbs.** Garlic braids, mushroom logs, potted plants, dried herbs, pepper wreaths, seed packets and grow-it-yourself kits, recipes with appropriate accompanying dried herbs and mushrooms, herbal dips, seasoning mixtures, relishes, mustards, canned vegetables, salsas, popcorn.

❖ **Health and beauty products.** Shampoos, soaps, lotions, foot baths, herbal teas.

❖ **Other ideas.** Oils, leather products, honey, candles, T-shirts with farm logo.

To add value, processing or manufacturing is usually, but not always, necessary. Craft projects, for instance, are generally done by hand at a table, rather than in a USDA kitchen. Processing can be something as simple as cracking and removing the shells of walnuts or pecans so the "meat" is exposed. Each step in processing adds time or effort, but will result in a higher-value end product. For example, whole walnuts sell for only 10 to 50 cents per pound, while walnut meats sell for $7 per pound. Specific varieties of hand-cracked walnuts yield 25 to 30 percent nut meats. There are machines to crack walnuts, which decrease time and labor, but they also decrease quality. Machine-cracked walnuts yield only about 7 percent nut meats.

Processing changes the form of a raw commodity. There are several reasons for processing, as follows.

1. **Ease of storage.** It is much easier, for example, to fit packages of sausage and pork chops into a freezer than a whole hog carcass.
2. **Convenience.** Most consumers would rather buy shelled pecans or boneless chicken than go to the effort of shelling or boning on their own — if the price is right.
3. **Longevity.** Canned fruits and vegetables will easily outlast their fresh counterparts, as do jams, jellies, and dried fruits. Most meats can be stored safely for 3 to 6 months in the freezer, but meat begins to lose quality after that. Cured meats can be frozen for only 1 or 2 months.

PRINCIPLE: *Plan to have products to sell year-round.*

Processing Food

You need no special permits for craft projects, and will not normally require much processing equipment. If you sell value-added food, however, you are probably going to need an approved kitchen with appropriate permits to manufacture it in. You must check health regulations — local, state, and USDA. Call your local health inspector and Department of Agriculture for more information. University extension offices often have information on adding value and kitchen requirements.

If you need a processing kitchen, your first step should be to find out what is available in your area. When starting, it is usually better to rent a facility from someone else, both to save money and to get an idea of what you will need. Check out churches, schools, restaurants, and nearby value-added farm businesses for a kitchen you could rent that will fulfill your needs. Your local Health Department or extension service may be able to give you some leads.

You will also need information on how to do the processing and labeling. The state agriculture department or extension may have a value-added program to help with labels. If you sell across state lines, there will also be federal guidelines that must be satisfied. Talk to other added-value farmers and call businesses that have similar products in grocery stores to get information on companies that supply labels, cans, bottles, shipping packages, and other processing supplies.

With meat, processing often runs afoul of USDA regulations. There are practically no small-scale processing plants for chickens, for example. One way around this is to sell the animal "live," then slaughter it for the customer. Check with your state's Department of Agriculture for rules and maximum numbers allowed. (In Missouri, each farm is allowed to sell 20,000 live chickens per year without a USDA inspection.)

In general, you can butcher fish, chickens, and rabbits at home. To sell pork, beef, lamb, or cabrito (goat), the meat must go through a state-inspected or USDA plant, but you can still sell direct if you sell the animals live.

Processing equipment for meat animals can be purchased from a variety of sources. Processing equipment might include fruit and wine presses, grinding mills for grain, poultry shears, knives, food choppers, and sausage stuffers, plus bags, egg cartons, and shrink wrappers. A number of companies carry rabbit- and poultry-processing equipment. (See the appendix for more information.)

Farmers today are lucky to make a 5 percent return on their farms when selling raw products like corn. Food processors consistently receive a 20 to 25 percent return on their investment. If you have done your research and priced your products correctly, add-on value will increase not just gross income, but net profits as well.

Twelve Ways to Sell Your Products

> **PRINCIPLE:** *Direct marketing is the profit equalizer for small family farms.*

There are a multitude of methods by which you can sell what your farm produces. Let's take a quick look at some of these. The first eight listed here are direct-marketing opportunities; the others may involve selling to retail outlets.

Selling to Friends and Neighbors

This is the best place for farmers new to direct marketing to start. You and your spouse should compile a list of about 100 people you know through work, clubs, church, and so on. Include people like your banker, your barber, your letter carrier, and relatives. You will use this list for contacts, and, if you continue to sell this way, as a mailing list to let people know when your products will be available.

This is an easy way to start direct sales, but it does have some drawbacks. It requires a lot of work to contact people to sell your products. Also, because these are your friends and neighbors, they may expect to get a lower price as a friendly bonus. They may also be slow to pay. You must remind them that for you to stay in business and your farm to be sustainable, you must make a profit. A good business deal is good for both parties.

Friends and neighbors may be more critical of your product. You must explain exactly what they are getting. Customers not familiar with farming frequently assume a 250-pound hog will produce 250 pounds of edible meat. Instead, a 250-pound hog will probably produce 150 to 170 pounds of cut and wrapped meat, including chops, hams, and sausage. If you explain to your customers what to expect before they buy the meat, it is more likely that you will keep a satisfied customer. Remember, an unsatisfied friend or neighbor can cause a lot of problems.

On my farm, I sell hogs at 80 cents a pound liveweight, which means about $100 for half of a hog, with the customer paying for the processing. Processing costs run between $35 and $45, depending on how the customer wants his meat cut and whether he wants any cured meat. Alternatively, I sell 80 percent lean sausage at a flat $2.50 per pound, and get about 100 pounds of sausage from a 225- to 250-pound hog. I sold all my pork this year in the midst of 10-cent hog prices at the local sale barn. Customers were willing to pay premium prices for quality meat. My pork is fresher and leaner, and my hogs were raised on pasture and not given any antibiotics or growth hormones. My pork comes from happy hogs — and my customers know it.

Farmers' markets provide a lot of customers in one place, which works well if you have a product they want.

Farmers' Markets

Farmers' markets are perfect for direct marketing. Consider selling at any farmers' markets within 1 hour's driving time. Your state Department of Agriculture should have a list of markets.

The biggest advantage of selling at a farmers' market is that you'll find lots of consumers in one place. It is much easier to sell a little bit of produce to a lot of customers than a lot of produce to a few customers. If you have something new and different from the rest of the market, provide taste samples to the customers. Recipes featuring your produce are also a good marketing strategy.

The major disadvantages to farmers' markets are cost and time. Time spent at the market is time not spent on the farm. When you figure all your expenses — production costs, booth fees, labor, and transportation — selling here may not be cost-effective. Count your pennies carefully.

Check out the market the season before you intend to sell there. See what the farmers are raising, and what they are not. Is there anything missing — yellow or purple snap beans, or heirloom tomatoes such as 'Brandywine'? Talk to the farmers and customers to see what is needed that you could raise. This is a good place to start.

If there is no farmers' market in your area, consider starting one. Evaluate the customer potential, determine the exhibitor (farmer) potential, and design a charter — all these are very important. Decide fees, dates, and hours, and investigate insurance and location. This entails a lot of work and requires community support. If you do start a farmer's market, holding special events at it — bake sales, festivals, concerts — can be an additional way to draw revenue. Finally, a farmers' market will require the same thing you do — advertising to let people know it is there.

Roadside Stands

These will usually be on or near your farm and can vary from a pickup bed, to a temporary shelter, to a shop. If your own property is off the beaten track, inquire at area businesses such as discount stores and gas stations about setting up a temporary stand on their property, or pool your produce with a farmer on a good road. Make your stand attractive, with bright signs and clear information.

Roadside stands bring customers to you at little expense. If you do well, you can develop this into a tidy little on-farm shop. With some advertising, it could attract people from miles away. Unfortunately, time is again a negative factor. Time spent in the stand may be worth less than at a farmers' market (think in terms of customer dollars per hour), unless you are on a well-traveled road with an easy exit. You will have to decide what the stand's hours will be. Will it be open daily, three times a week, or only on weekends? Will it be open all day, or just around rush hour, when customers are heading home? In some areas, the farmer leaves a can for money and lets customers help themselves. I'm not trusting enough for this approach, although it might be a good way of disposing of excess produce. Check insurance requirements to operate a roadside stand.

Community-Supported Agriculture (CSA)

This is a subscription service where the farmer signs up customers for a monthly or seasonal fee and agrees to deliver to them a percentage of the farm output for a season. CSAs are more than just a way to sell your produce: You form a definite relationship with your customer, and your customer shares in the production risk of your farm. By receiving the yearly fees up front, the farmer avoids borrowing money and paying interest, and has a paid market before he or she plants anything. The customers know how and where their food is grown, and receive fresh produce at a good price. Customers will also get a good lesson in the vagaries of weather!

Working a CSA is relatively simple. Most CSAs have a starting base of about thirty customers. The customer will pay, say, $300 for 7 months, and in turn receives a weekly delivery of his percentage of the produce — probably about 10 pounds per week. Customers may be allowed to work on the farm in peak labor times to reduce their cost. Some farms have customers pick up their produce on-farm, with a "rollover" table where a customer can remove produce he or she does not want and substitute something else. Some farms provide a yearly list of what they are planting; others seek input from the customers as to their preferences.

The main disadvantages of a CSA are determining a fair price and finding customers to sign up for it. Many people are afraid to

TRUE STORY

Sam and Elizabeth Smith grew organic produce for local markets and restaurants for many years on their Williamstown, Massachusetts, farm. They decided to turn their farm into a CSA because they wanted a closer tie with the local community. CSA members actually become part of the farm. The Smiths have apprentices from all over the country who live on the farm from spring until harvest to learn about organic farming and the CSA experience. The apprentices not only help tend the crops and animals but they also help pick the vegetables when members come to do their pickup. Members choose certain crops, such as strawberries, raspberries, cherry tomatoes, snap peas, string beans, kale, herbs, and flowers.

As separate items, the Smiths' Caretaker Farm raises organic lamb, pork, and beef. Honey and eggs are available for sale. A local baker has joined the operation and provides fresh healthy breads, rolls, and special desserts, as well as baking classes. The Smiths don't grow corn, but it's available to members thanks to a participating neighboring farm.

Caretaker Farm has become a thriving community of members who value local, fresh, organic produce and enjoy their weekly visits to the farm to pick up their share. A steering committee helps manage the membership administration. There are seasonal festivals, such as planting and harvesting potatoes and harvesting pumpkins, with square dances and potluck dinners. There is a pond where members can swim; there is a children's garden and classes for children in making various craft projects.

commit money before getting any product. Most (but not all) successful CSAs are near metropolitan areas with a large customer base. You don't want to start a CSA until you have a good knowledge of what you can raise each year, and what it costs to raise it.

Catalog Sales

Value-added products are required for catalog sales, which can be a natural outgrowth from a mailing list of customers. Starting a catalog requires a lot of thought and careful market research. Printing, packaging, and mailing the catalog — that's the easy part. The hard part is juggling supply and demand, making sure you have the necessary supplies on hand without spending too much money or creating a surplus. If you don't have what the customer wants when he wants it, you have lost a sale and possibly a customer. If you have thousands of unsold jelly jars, you have lost a business.

I recommend catalog sales only for those with several years of experience in both farming and direct marketing of value-added products. A catalog requires a large volume of production. Factoring in the weather, amounts of production, and projected customer returns calls for knowledge and experience. It might be wise to have another farmer close by who can supply you with extra produce if you run short — but it must be of a quality comparable to yours.

Shows and Fairs

A booth at trade shows and community events can promote your farm and increase your sales. These booths are a good way to tap into local markets, appealing both to brand-new customers and to people who have already heard about your Aunt Sally's Special Mustard Recipe but have never tried it. You might consider offering some kind of exhibition in trade for the booth — cooking, milking a goat, or demonstrating a craft (soapmaking, spinning, felting, for example). Have cards or flyers available with maps to show interested people how to find your farm.

Check with the chambers of commerce in nearby communities, and also with your local extension office, to see what events are available to you. Don't look just at traditional farm and craft shows. Consider everything, from ballooning events to food fairs, from music festivals to church gatherings, from association conferences to Living History events. If no such events exist in your area, consider hosting one yourself.

The next step up from local events is a booth at a major farm and craft show or a state fair. You must be careful about costs. Remember that you have to recover the booth fee (usually $50 for small shows and $250 or more for larger shows) before you make a profit. If you are in another city, you may have to recover transportation, lodging, and meal costs as well. I recommend attending a major show the year before you exhibit at it to assess the customer potential. Talk to entrepreneurs marketing similar products and ask what they think of the show. In the end, though, finding out if you will have enough customers to justify the expense can be done only by trying a show or two and keeping good records. Be warned that some shows will be great, outdoing all expectations — but some will be duds.

The "major leagues" of shows are the food festivals and gourmet food fairs, where booths can cost $2,000 or more, plus fines if you don't do exactly as instructed. These are not for faint-of-heart, low-volume producers, or those without a good cash reserve. You must be professional, knowledgeable, competent, and know your costs of production to the tenth of a cent. If a buyer asks what it costs you to deliver 100 six-packs of jam to 20 different stores, you have to be able to say, "Yes, I can do that for XX dollars and cents per case."

U-Pick Farms

U-picks are definitely a "people-person" business. All sizes, shapes, and ages of people will come to your farm, full of all kinds of problems, stories, and misconceptions about farmers. They will gladly share their experiences while you try to deal with five customers waiting to check out, a small child trampling plants in the field, and cars blocking the parking lot. If you are a "people person," though, and believe you can sell your visitors products that are

Some families make it an annual event to pick their own produce.

wholesome while providing them a fun experience, this can be a good way to market.

U-picks do away with the expense and time of harvesting your crop and hauling the produce to town. (Some farms pick part of their crop and sell it to hurried visitors at a higher price.) Once customers are on the farm, you have a great opportunity to sell them other products, too.

There are many other cost considerations, however. You need to check into the requirements for U-picks before you start: insurance, parking, bathroom facilities, a shady rest area, and so on. For information, talk to insurance companies, other U-pick farmers, and your local extension office. Check county regulations, as well.

Think, too, about the problems inherent in letting large numbers of people onto your farm — consider fence placement and security, for starters. Once again, it is helpful to do your research, and consult with other U-pick farmers and your extension service. You'll need to offer pointers to inexperienced pickers to reduce damage to your crops. There will always be some free "field tastings," but discourage massive consumption. Require parental supervision of children — you are not a baby-sitter (unless you want to set this up as an extra perk for a fee). Some farms require phone-in appointments, while others just have drive-bys. Phoning allows control of customer numbers with production, but requires more advertising until you acquire a regular clientele.

U-picks used to sell only by the container, but most now choose to sell by weight. Thus, accurate scales are essential.

It is best to have a large population center within 45 minutes' driving time, unless you have some innovative ways to draw people to your property. Earnie Bohner, at Persimmon Hill Farm in Lampe, Missouri, operates a U-pick that is well over an hour away from any major city. He's on a road that leads to a camp, however, and many parents driving down to see their kids stop by. He offers homemade muffins and fresh berry smoothies, plus his unique add-on-value berry and mushroom products (dried berries, mushroom sauce, and berry barbecue sauces). His special products and fun farm serve as a tourist magnet. He even has tour buses come. There is more information on attracting customers in the advertising section later in this chapter.

Food Circles

A food circle is sort of a cross between a CSA subscription farm and a cooperative. A limited group of farmers and others in town pool their produce — usually 1 farmer for each 10 nonfarm families is a good ratio. Members may pay a fee to join, and some circles require a deposit. A tightly structured food circle will have a distribution site (perhaps more than one) where produce is delivered weekly. Members buy the produce at the distribution site. A member who does not buy enough during a certain period (monthly or seasonally) may have to pay a fine. A loosely structured food circle may just have people get together on an informal basis to trade or barter their goods.

The major disadvantage of a food circle is that lots of people must participate in starting it up and keeping it going. A farmer will find a lot of his or her time spent on organizational matters. It does have an advantage over cooperatives in that a set volume of produce is not needed from a farmer, and an advantage over CSAs in that the customer does not have to contribute as much money up front; this makes it easier to attract customers. Farmers may not be free to set the price on their produce, however. Most food circles set a consistent price for all like produce, combine it, and return the farmer a percentage of money equal to the percentage he contributed.

Grocery and Health-Food Stores

These can be good outlets for your products, but you will be selling them wholesale, so you must move more product to make up for your reduced income and increased time and travel. While helpful, any wholesale selling should be a small portion of your total sales volume.

TRUE STORY

I was interested in getting into an alternative agriculture venture that would give me more return on my dollar than cattle. I decided to investigate raising catfish. I did my research carefully on raising them, then looked into marketing. I inquired at grocery stores in the nearest large city (about 30 minutes away) to see if they would have an interest in purchasing catfish. They did. I then checked with the Health Department to find out how I would have to arrange the fish for sale. I was told to take them to the stores in a cooler with a layer of ice, then a layer of fish.

When I finally had fish to market, I took them to town, arranged as the Health Department specified. I visited the first three stores that had indicated interest and every one of them turned me down. When the fourth store said no, I asked the manager why he did not want the fish. He said he could not use fish in ice — they had to be in bags. I checked with the Health Department again and was told that bags were fine, as long as they had a hole in the bottom and ice around them. Once I changed to this, all of the stores bought my fish. The lesson here is, ask questions to find out what the customer wants.

Stores may be hard to break into if they do not already carry local produce. Try to guarantee them certain amounts, packed the way they prefer — they may be more amenable. Of course, make sure you will be able to supply the amounts you promise. Consignment sales may be a way to get your products into the store. These also have the advantage of giving you a higher profit, but you will have to spend more time in arranging displays and accounting.

Restaurants

This again is not a retail market, but you may get prices slightly better than wholesale. Target specialty restaurants with staff chefs: They will be more interested in buying fresh produce. They also tend to be very particular. It is best to talk to the chefs before you grow, to find out what they want and what quantities they will need. Bring complimentary samples so they can experiment. Some chefs can be a bit temperamental. Tactfully encourage their purchases with one-sheet newsletters of information and quotes from food fairs and gourmet magazines. If you find a chef receptive to trying your produce, grow one or two new herbs or vegetables each season and give them as samples.

Some more tips on selling to restaurants:

❖ Make sure anything you supply to a restaurant is clean and attractive. Dirty produce is a quick way to lose a sale.
❖ Restaurants should be in a tourist area or close to a large population, or their volume may not be worth your time and effort.
❖ Try to have several restaurants as customers, and have produce that is used in a wide variety of dishes. If your sales are based on produce used only in the chef's house special, or in a single salad, you will be in trouble if that dish changes.
❖ Arrange delivery dates. If the chef can't make his specialty when he wants to, you may lose him as a customer. Leave your number with the restaurant for rush orders.

❖ Above all, make sure to supply the quantities a restaurant requests. If the chef asked for ten cases, don't bring six. Never promise what you can't deliver.

It is important to evaluate the volume of business available for each of your products. When I started my aquaculture business, I was selling trout and catfish to restaurants. I found though, that I was selling ten catfish for every trout, because there were only enough upscale restaurants in my area to use 25 pounds of trout per week. In contrast, I sold 50 to 100 pounds of catfish per week to several "home-style" restaurants, plus these fish sold much better in the grocery stores. I soon dropped the trout and raised only catfish.

Market Pools and Groups

Market pools are often retail. They are simply a group of farmers in an area who agree to refer sales to each other and promote each other's businesses. A farmer may refer a customer to someone else in the pool if he or she is currently out of what the customer wants or if the customer is looking for something just a little bit different. This system also allows the customer to have more choice. To go a step farther, the farmers may band together to sell their stock as a group at the sale barn, or to a company looking for larger numbers than a single farmer can provide. If this is done on a regular basis, it would be better to form a cooperative.

Market groups are usually regional or national associations that try to put together buyers and farmers. There are a number of these groups, particularly in the herb and flower businesses. The advantage is that the farmer does not have to do sales. The disadvantage is the commission that most of these groups charge. To find these groups, start by talking to some of the herb and flower associations and magazines listed in the appendix.

Cooperatives

Cooperatives are another way of getting someone else to do the marketing. These are especially good for farmers who dislike the effort of direct marketing or who have no "people skills." Cooperatives work by gathering together a group of small producers who are all raising the same product(s). They pool their produce to provide the volume required by major buyers. The selling is done by a co-op salesperson, freeing up the farmers for more production time. The disadvantage is that farmers will not get the retail price for their produce (although they may get a small premium for quality), and part of their profits are eaten up by co-op fees (including the salary of the salesperson). A member may also be in trouble if the co-op has contracts for X amount of produce and his or her yield is poor that year. Depending on co-op rules, there may be fines or other penalties.

There are all types of cooperatives. A co-op may be as small as eight to ten ranchers pooling calves to get a bunch of 100 for sale as first-time calvers, or it could be a group of farmers specializing in heirloom tomatoes for delis. Co-ops can also be quite big, such as the sugar beet cooperatives in the Dakotas, where farmers who join sign strict production contracts.

Investigate cooperatives carefully before joining. If you want to start one, do your research first, and talk to lots of other cooperatives. Even loose cooperatives need rules and regulations for members to abide by. It must be determined what percentage of sales goes to co-op salespeople and advertising, how administrative duties will be carried out, and how many growers will be allowed in the co-op. The new "New Generation" cooperatives set strict limits for numbers of members, differentiating them from older co-op models such as the Missouri Farmers' Association and rural electric co-ops. In a small co-op, make sure the other members are all people you trust and are willing to do business with.

Pricing Your Product

The price you put on your crops, livestock, and value-added products is what determines your income. People new to direct marketing often underprice their products and thus cheat themselves. You must keep in mind that you have a quality product (if you don't, you need to work on it until you do) and that you deserve compensation for it. You should be reimbursed for your direct expenses in raising it, the cost of your land and tools, your time in labor and management, and any costs involved in adding value and marketing it.

Accurate farm records are essential. You must know your costs of production before you can be reimbursed for them. If you do not know your costs of production, including labor, you cannot know the profit margin necessary to stay in business. As someone new to farming, your production records will not be as accurate as they will after 5 years of recording expenses, experiencing weather variations, and selling. Remember to adjust your prices as you gain experience.

Costs of production for crops include:

- Seed or plants
- Purchased fertilizers (including the costs of transportation for driving back and forth from your local horse stable with "free" fertilizer)
- Any additional equipment you need for the garden and plants (pots, starter soil, fluorescent lights, cloches)
- Gasoline
- Depreciation of machinery
- Labor

If you grow four major crops, split the dollars spent on fuel, repairs, and maintenance among them.

Costs of production for livestock include:

- Purchasing the stock
- Feed
- Veterinarian visits
- Medicine
- Fencing and shelter
- Labor

A meticulous accounting will include such things as the miles you drive to and from the sale barn. Time spent selling door to door or at farmers' markets should be included, as well as any advertising dollars you spend.

Once you have allowed for expenses, decide what profit to allow yourself. This is where you need to check your competitor's prices. What are similar products selling for at farmers' markets? in grocery stores? at nearby roadside stands? Is your product different — perhaps of more quality, all natural, or an heirloom variety? If so, price it accordingly. The customer will not think it is quality if you price it too cheaply. If you're not sure people will pay a high price for your product, give out samples to encourage interest. I used to sell 'Red Cup', an heirloom stuffing tomato, at a premium price to a small deli. The deli, in turn, sold it as a special, stuffed with tuna salad. Any variety of produce not found in stores should be sold at a higher price due to its rarity. This applies to meat and value-added products as well.

You may be tempted to underprice your produce to make some quick cash. This is a mistake. If, for example, you were to undercut your fellow farmers at a farmers' market, you would make them angry. Also, new customers will expect to see these prices every time, and you will get a reputation for low prices rather than for quality. Finally, it will cost you money in the long run. A drop in price requires a much greater increase in sales volume for the same net profit. By the way, if you have extra produce at the end of the day, take it home — your chickens and hogs will appreciate the treat.

As time goes by, you may want to consider raising your prices. In general, for a moderate price increase, determine if the total dollar increase in gross revenues is higher than losing the gross income of 10 percent of your customers (at least in the short term). Remember, it is better to raise the price in small increments over a long period of time than to have a big jump in a short period.

Advertising

PRINCIPLE: *Advertising is necessary to help your customers find you.*

Advertising is a great way to gather customers and increase profit at your farm, and it doesn't have to cost buckets of money. The first and best way to advertise is word of mouth. Tell your friends and neighbors to spread the word. Tell your barber or banker about your produce. Give out free samples at association, community, and church events. Make sure you include directions to your farm via business cards, brochures, maps, and recipe cards with your address included. Pool your resources with fellow farmers, and offer directions to each other's operations. Give talks or slide shows about your operation to local business associations, garden societies, 4-H and other children's clubs, churches, and schools — this is an excellent way to get free publicity. By creating a network of friends, relatives, neighbors, businesses, clubs, and community, you will enhance your business and your relationships.

The next step in advertising is flyers and notices. Place them on community bulletin boards, such as at hardware and grocery stores and local restaurants, and at service places such as barbershops and dentists' and doctors' offices. Local motels may place your literature on their display rack to show their lodgers things to do while in town. If you have seasonal or one-time events at your farm, you can always submit these as calendar items to your chamber of commerce, plus to local newspapers, magazines, and radio stations. Remember to allow lead time for publications before an event — some magazines may require two or more months' notice. Check with your extension office and state Department of Agriculture about any advertising opportunities they offer.

KEEP SIGNS SIMPLE

With roadside stands and farmers' markets, your advertising is often just words on posters or a wipe-off board. Make your signs attractive and colorful.

Roadside signs must be simple. A rule of thumb is to have 1 inch of letters for every 10 miles per hour a car will be traveling. On a 40-mile-per-hour road, make your letters 4 inches tall. Put the signs far enough before the turnoff that cars can slow down in order to turn. I recommend placing a sign on roads approaching your farm from each direction and another sign at the turnoff. Put the signs about ½ mile from your farm, clearly indicating the remaining distance.

Make signs in colors that stand out (yellow is not a good choice). On a colored background (such as neon pink), use only black lettering. The more professional and attractive the sign, the more business you will get.

Make sure your roadside signs are visible and easy to read.

Buying classified and display advertisements in ad gazettes, newspapers, and magazines may be useful, but it can also be expensive. This is even more true of radio and television. Instead, consider trying to get them to do a feature story on you. It is great publicity at no cost. Kelly Klober, direct marketer and author of *A Guide to Raising Pigs,* recommends sending a box of samples to local radio or television personalities to promote your farm.

Klober also recommends maintaining a mailing list of customers and sending them word of developments on your farm: new produce or livestock, baby animals, start of the season, special opportunities. When you have 200 names, investigate acquiring a bulk mailing permit at the post office. Make sure all the local media are on your mailing list. Send them regular press releases or a newsletter, or even just a postcard. Press releases should be limited to a single page, double-spaced.

GETTING THE MOST FROM YOUR AD

When you buy advertising in newspapers and magazines, there are some rules to consider. A classified ad should have a catchy headline (Fresh Wholesome Vegetables, or Lean Tasty Beef) and contact information. If you are providing a service such as delivery, list it.

Display advertising should have a catchy headline, a brief description of the product, contact information, and a picture — either a photograph or a line drawing — to get the customer's attention. Don't overload your ad with words or fill all the available space: White space will also draw the customer's eye.

Good placement also enhances an ad's sales potential. The right-hand outer column is the best place to be. Talk to the publication's salesperson about preferred placement. There may be a charge for the best position, or you may get it free as a bonus if you buy multiple ads.

Talk about prepayment and multiple-time discounts, and anything else you can do to bring the price down. Newspapers have fairly set prices, but magazines and ad gazettes may have some bargaining room.

Web sites are an up-and-coming sales technique. You may be able to arrange for notice on your community's site, or you may want your own Web page. Check out different companies for prices and services. It is best not to be buried too far in another company's Web site (e.g., http://www.freshlamb.com is a better site choice than http://www.somecompanies/htolg/findit/services/farms/freshlamb.com). Do not spend too much money on this method until you are sure of its effectiveness. On-line catalog sales may be something to consider as you expand.

Your Farm as a Destination

There are other ways to attract people to your farm, either by offering additional activities on the grounds or by providing services that keep customers interested in coming back. Activities might include fee hunting or fishing, swimming and boating, a bed-and-breakfast, ice skating, a petting zoo, a restaurant or café, a craft shop or art gallery (featuring works of local/regional artists), and a children's play area (hay-bale mazes are popular).

You can encourage community and farm associations to hold events at your farm. This is another opportunity to sell your products and develop more names for your mailing list. Anything from a balloon race start (but consider the effect on your livestock) to a conservation wildlife day may help draw customers. You might sponsor your own Farm Day with fun activities to attract townsfolk, or a Field Day with an explanation of your profitable farming methods to attract fellow farmers. You may want to have a value-added workshop or class, where you charge people to come to your farm. This allows you to market your skills as well as your product. If you develop a farm kitchen, rent it to other value-added agripreneurs.

Talk to schools, daycare centers, senior citizens' homes, travel agencies (touring companies), and clubs about doing tours of your farm. You can either conduct these free as promotional events or charge a fee per person or busload. Senior citizens enjoy visiting farms for reasons of nostalgia. They will all have a "snakes in the blackberries" story and a "Remember when . . . ." A few busloads of senior citizens at a flat fee per person (say $2), plus additional sales, will enhance your profit margin. Visitors will also be ripe for future sales such as Christmas gift packs, and will share their great experience with their friends and relatives. Make sure they leave the farm with a brochure or catalog.

Seasonal attractions such as hayrides and bonfires are another way to get people to come out. Decorate your farm for Halloween or Christmas. Sponsor a School Day through local elementary schools, where children get a small pumpkin free. (They'll encourage their parents to bring them back to the farm to buy bigger pumpkins and other sale items.) Have a pumpkin-carving contest (entry fee: just the cost of the pumpkin), with space donated for advertising through the local paper. Emphasizing a family farm experience helps make these visits an annual event, at the least. Sponsor a pumpkin or Christmas tree fund-raiser through a children's organization (4-H, Boy or Girl Scouts, for example) or a church or service group. Sell value-added holiday corn decorations or wreaths. Have an egg hunt at Easter, with sales of value-added decorated eggs or rabbits as pets. While you have the visitors on the farm, tell them about other opportunities

A welcoming farm on a well-traveled road will be popular with locals and tourists alike.

they will have, such as buying fresh sweet corn in the spring. Keep a sign-up sheet on hand to expand your mailing list. You can also acquire addresses off checks.

Of course, to get people to come back, you will have to offer attractive services. A graveled parking lot and either gravel or concrete walkways are a big plus. If your fields are not close to the parking, offer wagon or cart rides out and back. Put straw between rows on a U-pick farm, or even plant grass for a nice walkway. Strategically located trash cans are essential for your own benefit. Shaded rest or picnic areas, hot or chilled food and drinks (pies, bread, muffins, cider, fruit shakes, hot chocolate), nice rest rooms, soda machines, shopping bags, all increase your customer base and profit. Your operation needs to have an attractive appearance.

Here are a few more tips to keep in mind:

❖ Anyone manning the cashbox should be friendly, knowledgeable, and clean.
❖ Your produce should be attractive and clearly labeled with prices and types.
❖ Free samples always boost your sales, but be careful not to give away too many if you have large numbers of visitors.
❖ Accepting credit cards will almost always increase sales.
❖ Consider an 800 number for value-added sales.
❖ Always keep an eye out for what attracts you when you visit other operations and events. Innovation and an attractive operation will bring in customers — and keep them coming back.

❖❖❖
FOOD FOR THOUGHT
Many people consider marketing to be the hardest part of their operation. It's not easy — but direct marketing is what will improve your profit margin. Start planning your marketing approach now, including market identification, niche development, add-on value, and advertising, and your operation will prosper.

❖❖❖

Selecting Your Enterprise

Once you have set your goals, it is time to select your enterprise or enterprises. To do this, you'll have to research your farm and the enterprises you're interested in, then determine what you'll do and how you'll do it. Let's look at the factors you should consider:

❖ **What can your farm support?** Soil test your farm soil to determine what crops will grow best with its natural inherent fertility. Blueberries, for instance, like acid soil. Legumes such as clover and alfalfa prefer sweeter soils. Consider slopes and types of growth already present. A hilly area is usually more suited to livestock than to crops. Wooded areas might be left for tree crops.

❖ **How will you water your crops and livestock?** You can plan flood irrigation, soaker hoses, drip irrigation, sprinklers, or a bucket. Consider your water sources in terms of access and cost. If

◄ *Heirloom apples are in high demand today due to their superior flavor. They are just one of many possible enterprises you can choose to pursue.*

you choose the bucket, your water source had better be close to the crop or a water trough. You may have a well, a spring, ponds, a rural water district, or heavy rains. Make sure you save all the water that falls on the farm. (See chapter 5 for more on water.)

❖ **What are the costs and returns of your enterprise?** Before you begin, plot a cash-flow budget for the season, including harvest and storage costs. Make both short- and long-term projections. Consider legal requirements, equipment and labor, transportation, facilities, and start-up costs to acquire seed and livestock. Returns will depend on your production levels and marketing methods. Consider how much of a particular crop to grow, and whether you will keep it on a small scale or expand in the future.

❖ **Are there permits required or special regulations for your product?** Many states have special requirements for shipping nursery plants or reptiles in and out of the state. Others require a permit for certain operations, such as a release from the Conservation Department to raise deer or a permit for fee hunting for game birds. Processing kitchens must require fulfilling regulations of city and/or county health codes. Remember to add to your start-up costs fees for permits and money for equipment to satisfy regulations.

❖ **You will need to keep records.** Keep track of expenditures, livestock breeding schedules, crop yields, weather, and other markers you can use to improve your farm and choose future enterprises. Will you need any special records, like sales tax records and bloodline papers? Consider the time and cost of these records. To keep purebred stock, for example, you will probably want it registered, and may have to pay both for registration and to be a member of the registration association.

❖ **Can you plan in some quick returns for start-up?** Start-up capital is important. Once you know how much capital you need or can afford, determine how soon you can get a return on your investment. As I have said before, it is important to look at not just what you can raise, but how you will market it. Radishes are the fastest crop I know of. It takes 10 to 20 pounds of seed per acre to yield 900 to 2,000 pounds of radishes in only 21 days. This is all good, but can you market this many radishes from 1 acre — much less 5 acres or more? Ready-to-lay chicken pullets can be laying inside of a week of delivery. Snap beans take 50 to 60 days, while blueberries take 3 to 5 years before in full production. You need to consider a mix of short-and long-term investments.

❖ **Must you have general farm liability?** Will the enterprise need liability insurance? U-picks and roadside stands require insurance for injury to customers. You may want to check into crop protection insurance, although it is usually harder to secure at a reasonable cost for alternative crops.

❖ **When will you start?** Start-up time is extremely important. In traditional agriculture, you can look at data from the starting year forward and determine whether the farm will make it, depending on how many good years of livestock and crop prices followed the start-up year, and the debt involved. The wrong year to start or expand caused great financial stress to farmers in the drought years of 1953 and 1954, and during the rural crises of the 1980s, for example. Although alternative agriculture is not quite as tied to the climate as is traditional farming, choosing when to start is still important (see box, page 154). This is as true of farmers changing their operation as it is of those new to farming.

A new farmer I knew tried to start a vegetable operation in July. That particular year the weather was drier than normal, and with no practical way to get water, the start-up operation faltered. He become discouraged and quit farming. Patience goes a long way in any business — it is important not to rush your start. Research carefully and choose your time, when the season and weather are appropriate.

Starting Out

When should you start farming? The best time to start is the late summer — say, August or September before the growing season you want to be producing income in. If you cannot do this for some reason, grow something in your backyard or in a vacant lot, but gain experience. Working on someone else's farm is a good idea.

Assuming you can start in late fall of the season before you want or need the income, what do you do?

❖ Lay out a work plan for everything you have to do.

❖ You should already have the basic tools necessary for the coming year. Get them to your farm, and grease, oil, and otherwise prepare them for use now and storage over the winter. Buy whatever equipment you still need in the fall or before the spring season, when it costs more and you don't have time.

❖ Plan your marketing program. How much will you have to sell? How many customers will you need? Will you have to develop a market? Is the market already there?

❖ Lay out fields and buildings.

❖ Build whatever fencing and buildings you will need.

❖ Take soil tests.

❖ Gather as much free fertilizer as you can: Clean out barns, gather leaves and manure, start a compost heap, and spread it on crop ground.

❖ Plant fall cover crops so you have green-manure crops to turn under in the spring.

❖ Buy livestock as needed. Breed your animals to give birth in early spring.

❖ Order seeds and plants in January.

❖ Thirty days before the frost-free date (usually in April), start crops that require extra germination time (peppers, tomatoes, gourds, melons) under fluorescents lighting in a warm building. After about 2 weeks, take these transplants outside to a sheltered spot to harden them off, returning them inside at night.

❖ Plant everything on or after the frost-free date.

❖ Think about the big picture all the time.

❖ **How does the new enterprise fit with the rest of your farm projects?** You must consider what your enterprise will bring to the mix of crops and livestock that you already grow or want to grow. Rotating crops and livestock on your land will keep it healthy and rich. If customers will drive up to 50 miles for fresh berries, but will travel only 2 miles for snap beans, you may need berries on your farm in order to sell snap beans. Crops like pumpkins and Indian corn provide seasonal income, but are heavy feeders and will deplete soil nutrients. Season extenders like greenhouses enable you to have early and late crops, but they take up space and money. Tunnels or cloches may be an inexpensive alternative, but may also require more labor and management.

Choose crops and livestock that support each other and reduce the costs of their partner. A simple example is hogs and corn. Train the hogs to graze the cornstalks during late growth and turn the animals onto the field after harvest. Feed the corn to the hogs throughout the year, and the hogs will turn under the pasture and fertilize it.

❖ **What type of production method works best and is most profitable for this enterprise?** Consider conventional, sustainable, natural, and organic methods. Crops can be grown in greenhouses, raised beds, rows, or trellises. Livestock can be raised on feedlots or pastures. You may want to practice management-intensive grazing (see chapter 4). Carefully examine the costs of production, labor, yield, and market potential to determine how to proceed.

❖ **Will the enterprise provide seasonal or monthly income or some combination of the two?** If it provides a short-term income during the year, combine it with other enterprises whose incomes mature at different times. Consider adding value to extend your enterprise's income period.

Comparison of Farming Types

	Conventional	Sustainable	Natural	Organic
Equipment	Standardized, easy to find	Standardized, easy to find	Standardized, easy to find	Difficult to find
Pesticides	Acceptable	Low input	Only natural	Only natural
Antibiotics/ hormones	Acceptable	Not recommended	Not allowed	Prohibited
Fertilizers	Synthetic okay	Low input or natural	Synthetic okay	Only natural
Market	Established commodities markets	Largely direct markets	Commodity/ direct markets	Largely direct markets
Information access	Easy	Difficult	Difficult	Difficult
Profit margin	Low	High	High	Very high
Financial risk	High	Not as high	Not as high	Not as high
Weather risk	Weather dependent	Not as dependent	Not as dependent	Not as dependent

Note: Organic farming requires certification from an organic agency. Many farms have both conventional and natural systems, or natural and organic systems, working together as the farmer makes the transition to low-input, sustainable agriculture.

❖ **What are the labor requirements of your new enterprise?**
For example, berries require large amounts of hand labor. If the enterprise is constructed as a U-pick operation, though, with the customer doing the majority of the harvest labor, or as a CSA (community-supported agriculture; see pages 132–134), where shareholders share the risk and provide some labor on the farm, your labor will be reduced. Consider all of the labor for all of your crops and livestock, and leave some time for yourself and your family.

❖ **Who will furnish the labor for your enterprise?** It may be you alone, your family, or hired labor, seasonal or full time. Labor costs in time and money must be considered for planting, harvesting, and marketing, to determine how many people are needed for a particular crop or livestock.

❖ **What will be the machinery requirement for this enterprise?** Consider your alternatives in terms of both time saved and investment costs. Will a $500 rototiller do the job or will it take a $3,000 to $5,000 used tractor? Consider renting, leasing, and buying options. Check with your neighbors to see what they have available and what you can barter for. If you need specialized planting equipment or harvesting equipment, like a potato digger or a cherry tree shaker, make sure you can get it. There is harvest machinery for just about anything you want, but do you have the volume necessary to make machine-harvest a viable option?

❖ **What kind of industry support is available?** How accessible are the plants, seeds, and breeding stock for your enterprises? You may also need research or management support. Look for production budgets, and check for university or extension personnel with knowledge of your crop or animal. Check with area farmers and associations for expert consultants. If no research has been done on an alternative crop, start small and grow into it.

❖ **How will you market your products?** You can sell whole-sale, at a lower profit, or retail at a higher profit. You can add on value to a raw commodity to increase your profit and market time, at the cost of more work and investment. Contract selling locks in sales price, but many clauses for grades and quality are subject to discount. For the most part, you are better off doing it all on your own, and carrying all the risk but taking all the profit. Nobody can watch all the eggs in your basket better than you.

❖ **Where will you market your products?** Much as your farm risk is reduced by having a variety of crops and livestock, your profit risk is reduced by having a variety of markets. A farmers' market is a great place to sell produce, but you have to haul it to town. U-picks get the customers to provide part of the labor but require good liability insurance. Roadside stands provide good sales if they are located on a well-traveled hardtop road with room for parking. Route sales work well if you can combine the trip with some other project to lower your costs. At one time, when I had 200 laying hens, I delivered eggs to businesses along the road on the way to get feed every Friday morning. (See chapter 7 for more marketing information.)

❖ **Is there a market already established or will you have to create your own?** Even if there is a farmers' market in your area, it may not allow sales of your product. Some markets do not permit meat products or craft projects to be sold. Local stores may be able to buy from local farmers or their buying may be tied to an out-of-state headquarters.

❖ **How will you advertise?** You must advertise to let people know who you are and what you have for sale. You can use word of mouth, newspapers, magazines, radio, television, or direct mail. (See chapter 7 for more information on advertising.)

❖ **What level of experience do you have for this enterprise?** Are you willing to learn? Raising elk is similar to raising cattle, and could be an easy change. If you have never planted a crop, it is better to start with a garden than with 5 acres. If you have not practiced direct marketing, consider how much you like dealing with the public, and how good you could be as a salesperson. Which family members are best suited to deal with the public? The key is to be always friendly, be willing to learn from your customers, and never be afraid to ask questions.

❖ **How will you price your product?** You can sell by volume, by weight, by the head, or by the pound. The final price must include the percentage of profit you need plus all of your costs, including labor. Consider what the market will stand when planning your price. Can you sell your product above, below, or the same as store price? Do you have any special features to raise the price, such as organically grown, higher quality, or uniqueness? If you are the only person in your area with Christmas trees, you have a mini-monopoly and can sell at a premium price. As the only person selling okra, you can sell it at a premium price if there is a demand for it. If no one is familiar with okra, you will have to create a market before it will sell well.

❖ **How much competition is there for sales of your product?** Look at other farms engaged in your enterprise within 25, 50, and 100 miles. Look at how they are marketing it and what volume of sales they have. Can you offer something they don't? If you and forty other farmers are selling a standard breed of tomato at midseason at a local farmers' market, count on prices no better than wholesale.

❖ **How do you feel about your enterprise?** There's an old saying among cattle buyers: "Bought right is half sold." The same is true in picking a business enterprise. Never choose an enterprise just because it will be profitable. You have to like what you're doing; you'll be more successful if you enjoy your work.

The items I have listed here are all questions you must ask yourself about any new enterprise. However, they are not the only questions. In researching your choice, ask yourself whatever questions arise from its unique circumstances. Keep in mind its requirements for planting, growing, harvesting, and marketing. By approaching the planning in a systematic way, you will find out not just about this one enterprise but also how it compares to other enterprises. This way, you'll be able to have the most profitable mix on your farm.

Types of Enterprises

What you can raise makes an almost infinite list. It expands even more when you factor in how you are marketing your enterprise. I want to cover a few enterprise choices, along with some of their advantages and disadvantages and their potential markets.

❖ **Cattle.** Cattle are normally raised for beef or dairy. Both require large volume, unless you can develop direct markets. *Consider* small breeds such as Dexter cattle and Miniature Herefords, less common breeds such as Murray Grey and Belted Galloways, intensive grazing, raising organically or naturally, selling beef halves or quarters.

❖ **Hogs.** To be successful, aim for the purebred or value-added markets. *Consider* rare breeds such as Mulefoot and Tamworth, raising on pasture, organic or natural, selling meat cuts or sausage.

❖ **Sheep.** Traditionally, sheep have been raised for wool, but there are increasing meat and dairy markets. The wool market is not profitable unless you add value through spinning or weaving. Some sheep are being raised exclusively for meat, such as Katahdin Hair sheep. There is some trophy hunting being done with breeds like Barbados Blackbelly. Dairy sheep like East Friesian are being used for cheesemaking. *Consider* rare breeds such as Navajo-Churro and Miniature Babydoll Southdowns, "new" breeds such as Dorper (a meat sheep), intensive grazing, selling meat cuts or sausage, seasonal lamb markets.

❖ **Goats.** Meat and dairy goats are common. Some fiber goats, such as angora, are prized for spinning and weaving. *Consider* ethnic markets for cabrito (goat meat), crossbreeding with Boer goats (a prized but expensive meat goat from South Africa), goat soaps, miniature breeds such as Nigerian Dwarf.

Dairy goat

❖ **Rabbits.** Rabbits are raised for meat and pelts. Small and dwarf breeds are sold as pets. Some breeds, such as angora, are raised for fiber for yarns. Deborah Lemmer, a canny marketer from Shaggy Shagbark Acres in Michigan, likes to demonstrate spinning directly from a rabbit sitting quietly in her lap. *Consider* rabbit jerky and sausage, restaurant sales.

❖ **Poultry.** Poultry includes chickens, turkeys, ducks, guineas, geese, and peafowl. There is a large chicken fancier market. You can sell purebred poultry, chicks, eggs, or meat. One farmer in Canada raises chickens bred exclusively for fly-tying feathers. *Consider* rare breeds such as Dominique chickens and Khaki Campbell ducks, pastured poultry, egg ornaments, ethnic markets for black-feathered breeds, organic eggs.

Ducks

Turkey

❖ **Elk, deer, and bison.** These are gaining acceptance in the meat market for being low in fat and high in protein. All require a large initial investment in fencing and permits. Elk are now considered livestock in many states. Velvet from elk and deer antlers is a high-ticket item used for medicinal purposes. The purebred market is doing well selling breeding stock. Hides are also a good seller. *Consider* jerky and sausage, restaurant sales, trophy hunting.

❖ **Ratites.** The bottom dropped out of the speculation market on ostriches, emus, and rheas a few years ago, but they still have sales potential. Meat on all of them is low in fat and high in nutrients. Hides and eggs have value-added potential; ostrich feathers and emu oil are also possible products. Initial investment is fairly low, but it takes effort to build a market. *Consider* sales to gourmet restaurants, emu oil pain rub, decorations with hide and feathers, carved or painted eggs.

❖ **Game birds.** Pheasants and quail can be raised for breeding, meat, or hunting. Initial investment is fairly low, but permits are needed and markets must be sought. *Consider* sales to restaurants and upscale grocery stores, selling birds to hunters or providing hunts on your property, catalog sales of chicks.

Llama

Alpaca

❖ **Camelids.** Llamas and alpacas are used primarily for fiber and breeding markets. Llamas are gaining acceptance as pack animals and guard animals for sheep flocks. There is really no meat market for camelids. Alpacas can be a fairly hefty investment, but items made from alpaca wool sell at a premium. Heat can be a problem. *Consider* Alpaca yarn, llama sweaters, blankets, guided trips in National Forests with llamas.

❖ **Small mammals and reptiles.** Small mammals include African Pygmy hedgehogs, Sugar gliders, chinchillas, and gerbils for the pet market. Costs are low and space requirements are few, but permits may be required. Locating a market is the hardest part. *Consider* pet shows and sales, pet stores, local fairs.

A FISH STORY

There are twenty-eight major fish-producing countries. On a world-wide basis, at least ninety species of fish, thirteen species of shrimp, prawns, and crayfish, and a wide variety of shellfish and marine plants are produced using aquaculture techniques. Many developing countries are turning to fish farming to satisfy their protein needs, increase their self-dependence, and reduce risks from the dwindling ocean fish supply.

According to the Global Aquaculture Association, as reported in the January/February 1998 issue of *Fish Farming News,* trade in seafood products now generates $50 billion annually. There is an increasing demand among Americans for fish products. Shrimp accounts for about 25 percent of the fresh and frozen seafood consumed in the United States — and farm-raised shrimp makes up nearly one-third of the total shrimp supply. Population growth is expected to increase shrimp demand by about 2 to 3 percent per year. Currently, the USDA spends only about $\frac{1}{25}$ as many dollars on aquaculture as it does on crop research, but that is increasing.

The potential of aquaculture for a farming enterprise or home consumption is tremendous. I raised catfish in the 1980s in cages 36 inches in diameter by 4 feet deep. My already existing ponds were too deep to seine; the cages made them usable. It was fairly easy to produce 400 to 600 pounds of catfish in one of these cages in just 20 or 21 weeks, with feed efficiency of 1½ to 2 pounds of feed per pound of gain. A 5 percent death loss and a 55 percent dress-out (the percentage of usable meat after the removal of skin and viscera) grosses about $500 per cage, or could furnish about 210 pounds of fresh fish for your freezer. I experimented the first year, then built a 10-foot by 6-foot fish-processing plant complete with a used 6-foot by 6-foot walk-in cooler, and butchered my own fish as well as neighboring farmers' fish.

❖ **Aquaculture.** Catfish, trout, and bass can be raised for breeding stock, meat, or fee fishing. Meat fish such as tilapia are used in recirculating systems. Exotic small fish can be raised for the pet market. Costs can range from very low to extremely high. Farm-raised catfish and trout have good grocery and restaurant sales potential, although a small processing kitchen will be necessary. *Consider* restaurant and grocery store sales, fee fishing, sales at farmers' markets — live (if allowed) or packed in ice — sales through CSAs (see chapter 7).

❖ **Vermiculture (worms).** Selling worms for bait or for farm fields has a lucrative market. Some are grown and sold in compost kits for city dwellers. Low cost, but it takes work to keep selling them. *Consider* selling with "compost kits" at environmental shows and Earth Day fairs. Sell kits to local schools.

❖ **Bees, butterflies, and other insects.** Bees can be sold for breeding or rented for pollinating crops. Their primary market is value-added products from pollen, wax, and honey. Butterflies for pollinating and release at events such as weddings are gaining in popularity. Beneficial insects are also gaining acceptance as chemical use declines. These are all management-intensive, and market potential may be small. *Consider* grocery store honey sales through a grocery store or CSA (see chapter 7), trading pollination for a crop share.

❖ **Dogs.** Dogs are raised for pets, breeding stock, livestock guarding, herding, and hunting. The market is well established and steady. There are hundreds of breeds to choose from, and investment cost is low. If training is required, dogs are obviously management-intensive. *Consider* training dogs, stock dog clinics and shows, dog shows.

❖ **Equines.** Horses, donkeys, and mules are not inexpensive. Draft breeds are gaining in interest, but require a lot of time and effort to raise and train. Riding stock will not sell easily unless you are well known, although you might consider trail riding or establishing a stable. Donkeys are being used as livestock guardians. *Consider* breeding, racing, draft animal schools, Plow Days (a field day of draft-animal demonstrations to gather customers).

❖ **Pasture, hay, and cover crops.** These are the perfect way to combine crops and livestock. Except for baled hay, their market potential is not high, but they are necessary components in the interaction of different enterprises on your farm. *Consider* hay trade to neighbors for crops, hay-bale mazes at a U-pick, miniature bales (6 inches long) for sale at country craft events.

❖ **Traditional grains.** It is hard to profit from field corn, wheat, and soybeans without adding value. There is excellent potential for open-pollinated corn and unique soybeans, such as tofu. Open-pollinated corn is desired for its high feed value (usually more nutritious than hybrid corn) and its open-pollinated qualities, specifically, the ability to save seed from it for replanting. Some breeds also have sales potential for decorative purposes ("Indian" corns and Cherokee Blue, for example). Less traditional grains such as barley, oats, and sunflowers may have market potential, but should be considered mostly for on-farm use. *Consider* cornstalks and miniature corn for Halloween, chocolate-covered soybeans, organic markets, seed sales, cornhusk- and wheat-weaving craft projects, flours, and breads.

❖ **Vegetables and melons.** The traditional crops for direct marketing. A huge variety of both open-pollinated and hybrid varieties are available. Invest the effort to have them available before or after the regular season to increase sales. *Consider* rare and exotic varieties with a unique appearance or flavor, ethnic specialties, "baby" vegetables, pepper sauces and wreaths, tomato sauces.

❖ **Herbs and medicinal herbs.** Herbs are intensive crops with increasing market potential. Find buyers for medicinal herbs before raising them. High-dollar herbs, such as ginseng and echinacea, require a big investment in care. Herbs' best potential is probably on a small scale, as accompaniment to value-added items and farmers' market sales. *Consider* value-added soup mixes, vinegars, essential oils, and soaps, or sell fresh specialty herbs like lemongrass to specialty restaurants.

❖ **Mushrooms.** Mushrooms are easy to raise but require a few years before realizing much income. To raise them on a large scale may require special facilities and a good source of wood. Sales of specialty varieties to restaurants, grocery stores, and consumers is lucrative, as is selling mushroom-growing kits — prepared logs inoculated with mushroom spawn. *Consider* value-added items like dried mushrooms and soup mixes.

❖ **Gourds and pumpkins.** These are good seasonal items. It is best to add value to them for marketing. *Consider* carved pumpkins, specialty miniature and white pumpkins, painted gourds, pumpkin pie (usually made from butternut squash), specialty gourds such as luffa and dipper gourds.

❖ **Industrial crops.** Crops such as crambe, meadowfoam, guayule, and jojoba are used for nonfood industrial applications, such as lubricants and plastics. These often have excellent potential for profits, but require an enormous investment in time, labor, and learning, without an obvious market readily available. I do not consider them sustainable at the current time, in that you cannot use them on your own farm, and you are at the mercy of a very small number of buyers for market price and availability. (If only one company will buy a crop and it goes belly-up, you could be in trouble.)

❖ **Small fruits.** Strawberries, brambles (e.g., raspberries), ribes (e.g., gooseberries), and other berries have great potential for U-pick and value-added markets. Planting varieties that mature at different times of the season will ensure a steady flow of customers. Grapes have value-added potential as juice, wines, and flavoring. Unusual varieties can attract interested customers. *Consider*: U-picks, farmers' markets, grocery stores, health stores, syrups, dried berries, chocolate-covered berries, jams and jellies, fruit drinks.

❖ **Fruit and nut trees.** Orchard fruits like apples, pears, peaches, and cherries, and nuts such as hazelnuts, walnuts, and pecans do well at U-pick farms and farmers' markets, and may have grocery potential. For fruits, particularly apples, try to have varieties with different flavors and that mature at different times. *Consider* value-added potential, or a "fruit and nut subscription service," where a basket is mailed to the customer two or three times a year.

❖ **Flowers, cut and dried.** The cut-flower market is growing (pardon the pun), and there are certainly a multitude of varieties. With a greenhouse and a careful mix of maturity times and types, sales can continue year-round. Combine it with a nursery tree operation. There are good value-added markets (get-well baskets, dried arrangements) and also seasonal potential (Valentine's Day, Easter). *Consider* craft show sales, grocery stores and farmers' markets, flower shops, providing arrangements to upscale banquet halls.

❖ **Timber.** Tree crops can be raised for timber, firewood, or nursery stock. Sell nursery stock for windbreaks or wildlife shelter. When planted carefully and allowed to grow to maximum potential, tree crops can be highly profitable. One good choice is a quick-growing variety, such as *Pawlonia* or bamboo. Christmas trees also have excellent seasonal sales potential. Think before you clear an area of trees — you may want to call your extension service or Forestry Department to get an evaluation.

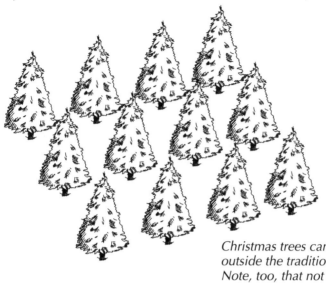

Christmas trees can provide income outside the traditional growing season. Note, too, that not all farm crops need to be edible.

❖ **Other alternatives.** There are many other alternative crops, from fava beans to shrimp, and from lions to sweet sorghum. The rarer it is, the harder it may be to locate a market — but then competition is reduced, too. Try to pick crops and livestock that will be easy to raise, can be marketed in two or three different ways, and support each other.

Enterprise Cost Analysis

Following are a list of budgets compiled from university, government, and private sources. These budgets are intended to help you in the process of selecting an enterprise. Keep in mind that you need to select those enterprises that match your sales, labor, and capital resources.

These budgets are good for planning, and for comparing "apples to apples" as to which enterprises are the highest grossing or most labor intensive. They also allow you to get a basic feel for investment and the labor involved. Let's analyze the budget below for a sow and sixteen pigs, direct-market as sausage, to see what information we can gather.

First, consider feed costs. The $1,116.00 feed costs divided by the gross income of the fed hogs, $4,687.50, means your feed costs are 24 percent of the gross income. Can you grow the feed or will you have to buy it? Will changing feed prices affect your project? In this case, prices will not have much effect, because the percentage of feed costs to gross income is relatively low and your profit margin is high.

What about labor? Half the labor, 20 hours, went to producing the product (hogs for sausage) and half the labor went to marketing the product (sausage). If we divide the net profit, $4039.50, by 40 hours, we find you have an income of $100.99 per hour for your labor expended on hogs.

Leaving out weekends and holidays, farmers need about 300 ten-hour days of labor to be fully employed, or 3,000 hours per year. (Yes, I know it's not fair that farmers have to work 10-hour days instead of 8, but that's part of the joy of being a farmer.) So using these figures, you would know that if you had six sows, ninety market hogs, and six cull sows, they would take 240 hours (6 sows x 40 hours) of labor for your hog project per year, leaving 2,760 hours for crops or other livestock projects. Dollars earned per hour is one way to measure profitability.

Another way to look at and compare profitability is net profit divided by gross income. In this budget, that equals 78 percent. A high percentage like this means that operating costs are low, as compared to projects like beef cows and stock calves, where the percentage is 61 percent.

Remember always that profitability depends on how well you market your product. Just selling retail instead of wholesale can make some projects profitable. In this hog example, it was profitable even with hogs selling for 10 cents per pound on the commodity market, because the product was not sold as a hog, but instead was direct-marketed straight to the consumer as a different product, whole hog sausage.

Profitability Comparison of Enterprises

Livestock

Beef cow/stocker calf

Income

500-lb. calf x .95 (calf crop) x .84 x $0.80	$319.20
One cull cow: 1000 lbs. x .16 x $0.40	$64.00
Total gross income:	$383.20

Expenses

Feed: hay 1.5 tons	$75.00
Pasture 3 tons	$38.00
Protein and salt	$5.00
Other var. costs	$30.00

Net profit $235.20

Labor 7 hours

Bison (1 cow)

Based on 15 cows and 1 bull

Gross income	$1,686.00
Expenses	
Feed	$486.00
Other var. costs	$155.00
Net profit	$1,045.00
Labor 10 hours	

Bobwhite quail (meat birds)

Purchased as day old, sold at 13 weeks, 3 groups of 1,000 birds, 15% death loss

Gross income 2,550 birds x $2.80	$6,757.00
Expenses	
Feed 9,112 lbs.	$1,457.00
Other var. costs	$1,680.00
Net profit	$3,620.00
Labor 220 hours	

Chickens, broilers

1000 chicks, 10% mortality, 8–9 weeks to market
900 chicks after mortality x 4 lbs = 3,600 lbs.

Income	
3600 lbs. x $2	$7,200.00
Expenses	
Feed: 10,800 lbs. (3 lbs. per lb. sold) @ $0.14	$1,512.00
Other var. costs	$1,000.00
Net profit	$4,688.00
Labor 500 hours	

Chickens, purebred

20 hens with production 60% (220 eggs per hen)

220 eggs x .60 (hatching rate)	132 eggs per hen
132 eggs x 20 hens	2,640 chicks
Income	
2640 chicks x $3	$7,920.00
Expenses	
Feed	280.00
Other var. costs	200.00
Net profit	$7,440.00
Labor 40 hours	

Dairy goat

Does averaging two kids

Income

    Milk sales, culls, and replacements $586.00

Expenses

    Feed $234.00

    Other var. costs $127.50

Net profit $224.50

Labor 22 hours

Earthworms

Production 4 pounds per square foot of worm bed

Income

    840 pints (100 worms to a pint) $840.00

Expenses

    Variable costs $80.00

Net profit $760.00

Labor 112 hours

Elk

Based on 25 cows and 2 bulls in the breeding herd

Income

    Breeding stock and velvet $4,144.00

Expenses

    Feed $429.00

    Other var. cost $149.00

Net profit $3,566.00

Labor 9 hours

Fallow deer (1 doe)

Based on 100 does and 5 bucks

Income

    Venison, breeding stock, hides $253.00

Expenses

    Feed $59.00

    Other var. costs $63.00

Net profit $131.00

Labor 3 hours

Hogs, feeder pig purchased: 40-lb. pig fed to 250 pounds

Direct market as sausage ($2.50/lb.)

Income

100 pounds sausage x $2.50	$250.00

Expenses

Feed	$48.00
Pig cost	$32.00

Net profit $170.00

Labor 8 hours

Hogs, sow and 16 pigs raised (2 litters per year)

Direct Market as Half Carcasses

Income

15 market hogs x 250 lbs. x $0.80	$3,000.00
1 cull sow (400 lbs. x .50 x $2.50)	$500.00
Total gross income	$3,500.00

Expenses

Feed	$1,116.00
Other var. costs	32.00

Net profit $2,352.00

Labor 20 hours

Hogs, sow and 16 pigs raised (2 litters per year)

Direct Market as Sausage ($2.50 per pound)

Income

15 market hogs x 250 lbs. x .50 x $2.50	$4,687.50
1 cull sow (400 lbs. x .50 x $2.50)	$500.00
Total gross income	$5,187.50

Expenses

Feed	$1,116.00
Other var. costs	32.00

Net profit $4,039.50

Labor 20 hours

Market labor 20 hours

Honeybees

Income
     60 pounds honey per hive x $3 $180.00
Expenses
     Variable costs $90.00
Net profit $90.00
Labor 11 hours

Pheasant

Purchased 200 day-olds, sold at 20 weeks for flight birds
Income
     80 roasters/80 hens $1,120.00
Expenses
     Feed $423.00
     Chicks ($0.90) $180.00
     Other var. costs $205.00
Net profit $312.00
Labor 120 hours

Rabbits (fryers)

20 does, 2 bucks, 5 litters per year
Sell 5 lbs. at 10 weeks
Gross income 3,500 lbs. x $0.80 $2,800.00
Expenses
     Feed 5 tons $1,400.00
     Other var. costs $129.00
Net profit $1,271.00
Labor 200 hours

Sheep

One ewe spring lamb born March, April, and May

Lambs 150% lamb crop, 20% replacements

Income

1.50 x 120 lbs. x $0.80	$144.00
150 lbs. cull ewe x $0.20 = 30 lbs. x $0.20	$6.00
Wool: 10 lbs. x $1.50 (added-value wool)	$15.00
Gross income:	$165.00

Expenses

Feed 500 lbs. hay x $.05/lb.	$25.00
Grain 100 lbs. ewe and lamb x $.035/lb.	$3.50
Fair pasture .75 acres	$11.25
Wormers, salt, minerals	$5.00
Other var. costs	$30.00

Net profit $90.25

Labor 6 hours

Sheep, feeder lamb

Bought 70 pounds, fattened to 120 pounds

Income

Lamb 120 lbs. x $0.75	$90.00

Expenses

Feeder lamb 70 lbs. x $0.78	$54.60
Feed	$14.00
Variable costs	$7.38

Net profit $14.02

Labor 1 hour

Sheep, meat type

One ewe spring lamb born March, April, May, 20% replacements

Income

Lambs 1.75 lamb crop x 120 x $0.70	$147.00

Expenses

150 x .20 = 30 lbs. x $0.20	$6.00
Feed	$44.75
Other var. costs	$30.00

Net profit $66.25

Labor 6 hours

Sheep, milk

Milking 160 days/year (lambing 1.5 times/year) 1.65 lambs per ewe

Income

160 lbs. milk	$104.00
95 lbs. lamb	$128.25
Wool and hides	$10.00
Cull ewes and ram	$7.50
Gross income	$249.75

Expenses

Feed	$85.45
Other var. costs	$20.86
Net profit	$143.44

Labor 15 hours per ewe

Crops

Asparagus

Average production 1, 2, and 3 years

Gross income/acre	$3,333.00
Total var. costs	$1,210.00
Net profit	$2,123.00

Labor 24 hours

Bell pepper

Gross income/acre (9,960 lbs.)	$3,357.00
Total var. costs	$2,334.00
Net profit	$1,023.00

Labor 185 hours

Blackberries

Years 2–15

Gross income/acre (8,400 lbs.)	$6,300.00
Total var. costs	$1,454.00
Net profit	$4,846.00

Labor 210 hours

Blueberries

Years 3–15	
Gross income/acre (4,000 lbs.)	$6,400.00
Total var. costs	$1,200.00
Net profit	$5,200.00
Labor 200 hours	

Broccoli

Gross income/acre (8,250 lbs.)	$4,169.00
Total var. costs	$1,002.00
Net profit	$2,136.00
Labor 136 hours	

Cabbage

Gross income/acre (16,250 lbs.)	$1,746.00
Total var. costs	$1,897.00
Net profit	$–151.00
Labor 115 hours	

Cantaloupe

Gross income/acre (10,000 lbs.)	$1,955.00
Total var. costs	$1,056.00
Net profit	$896.00
Labor 108 hours	

Cucumber

Gross income/acre (12,500 lbs.)	$2,150.00
Total var. costs	$1,255.00
Net profit	$895.00
Labor 153 hours	

Eggplant

Gross income/acre (16,500 lbs.)	$3,973.00
Total var. costs	$2,118.00
Net profit	$1,855.00
Labor 207 hours	

Lima beans

Gross income/acre (3,200 lbs.)	$1,600.00
Total var. costs	$985.00
Net profit	$615.00
Labor 128 hours	

Okra

Gross income/acre (9,000 lbs.)	$6,160.00
Total var. costs	$2,197.00
Net profit	$3,963.00
Labor 301 hours	

Onion

Gross income/acre (1,500 sacks 50 lbs. apiece)	$10,500.00
Total var. costs	$5,992.00
Net profit	$4,508.00
Labor 20 hours	

Potatoes

Gross income/acre (15,000 lbs.)	$2,150.00
Total var. costs	$1,086.00
Net profit	$1,064.00
Labor 105 hours	

Pumpkin

Gross income/acre (28,000 lbs.)	$2,800.00
Total var. costs	$1,419.00
Net profit	$1,381.00
Labor 25 hours	

Raspberries (Red Heritage fall berries)

Years 2–15	
Gross income/acre (7,500 lbs.)	$6,375.00
Total var. costs	$1,192.00
Net profit	$5,183.00
Labor 111 hours	

Snap beans

Gross income/acre (4,200 lbs.)	$2,163.00
Total var. costs	$1,002.00
Net profit	$1,162.00
Labor 109 hours	

Spinach

Gross income/acre (6,000 lbs.)	$2,220.00
Total var. costs	$1,031.00
Net profit	$1,189.00
Labor 103 hours	

Strawberries

Years 2–4

Gross income/acre (6,900 lbs.)	$4,116.00
Total var. costs	$1,466.00
Net profit	$2,650.00
Labor 96 hours	

Summer squash

Gross income/acre (13,860 lbs.)	$3,811.00
Total var. costs	$1,311.00
Net profit	$2,500.00
Labor 140 hours	

Sweet corn

Gross income/acre (900 dozen)	$1,363.00
Total var. costs	$705.00
Net profit	$658.00
Labor 71 hours	

Sweet potatoes

Gross income/acre (15,000 lbs.)	$3,350.00
Total var. costs	$1,570.00
Net profit	$1,780.00
Labor 115 hours	

Tomatoes

Gross income/acre (20,000 lbs.)	$7,686.00
Total var. costs	$3,101.00
Net profit	$4,585.00
Labor 306 hours	

Watermelon

Gross income/acre (17,000 lbs.)	$1,360.00
Total var. costs	$623.00
Net profit	$737.00
Labor 76 hours	

Diversity

PRINCIPLE: *A diversified farm has the best chance of success.*

Diversity in enterprises provides you with economic stability. By diversity I mean raising both crops and livestock, and more than one kind of each. If you have crops and livestock that you can sell at different times of the year, and have something else to sell when the market for any particular product drops, you will have a steadier income.

Enterprises that feed into one another strengthen this bond. Use rabbit waste to raise worms, or set up a recirculating water system to feed fish waste to plants. Running and feeding livestock on a plot that you will raise vegetables on next year can provide all or part of the fertility needed for your vegetable crops. Rotating your crops and using cover crops in between can also provide some grazing and possibly hay for your livestock. By using crop ground or vegetable ground for livestock in winter and crops in spring and summer, you can triple your gross per square foot.

This same multiplication factor works if you can graze cattle, sheep, and goats on the same pasture. This, however, is highly management-intensive and should be done only after you have gained some experience in intensively grazing one species.

A diversified, sustainable farm will give you the best chance to succeed in agriculture. Keep diversity in mind when choosing your enterprises.

Sustainability

> PRINCIPLE: *To be sustainable, farm enterprises must be profitable, environmentally sound, and socially acceptable.*

Sustainability on the farm means that it is profitable — an unprofitable farm cannot sustain its financial operations. If it is unprofitable, it must be subsidized with off-farm income. This is fine for getting started, and certainly cheaper than a bank loan. In the future, however, it should stand on its own — if the farm is not profitable, you will not be able to realize your lifestyle or monetary goals.

The enterprises that make up your farm must be environmentally sound and socially acceptable to be sustainable. Sustainable crops and livestock can adjust to and thrive on a low-input-type system. My Katahdin Hair sheep are a breed noted for their natural parasite resistance (lower worming expense). I lamb them on pasture in May when grass is up and growing. Katahdins have few lambing problems, and are good mothers and milkers. Under my feeding and grazing system — all grass, except when breeding them or during lambing season — I have little management or cost inputs. They provide me with steady income, year after year, at little cost to me, my farm, or the environment.

I believe free-range poultry and pasture hogs will once again become the norm in agricultural production due to their sustainable properties — low inputs, easy management, and environmental soundness. The other reason that they will do well is that today's consumer wants to know that his or her food is safe. Consumers are interested in who grows their food and how it is grown. Last, they are interested in quality. Consider the sustainability of any enterprises you choose, and how you can increase your sustainability over time.

❖❖❖

FOOD FOR THOUGHT

Now that you have this chapter under your belt, you should be able to pick the combination of livestock and crop enterprises that matches your farm, labor, and capital resources. If you choose carefully, you are on the road to success.

❖❖❖

PART IV

Management

Machinery

> **PRINCIPLE:** *Machinery and tools should save time or reduce the need for additional labor. Otherwise, they are a waste of money.*

Machinery, by my definition, is any piece of equipment or tool — including hand tools — that makes the job easier, faster, and better. A hoe, a wheelhoe, a two-wheel tiller, and a tractor are all machinery, each with advantages and disadvantages.

The purpose of any tool or piece of machinery is to save time by reducing labor. This increases the work that can be accomplished and decreases the cost of doing the job. To decide if a tool or piece of machinery is worthwhile, its cost must be balanced against the time saved and what you can do with that time. If a piece of machinery will not save you time or reduce the need for additional labor, don't buy it. If a tool saves time, but you don't use the time on other projects, that tool is unnecessary.

As I am a four-shot-a-day diabetic with three congestive heart failures, I use machinery because it saves the parts of my body that still work. I am also interested in producing a surplus of my crops and livestock so I have plenty to sell to my customers in a timely manner.

≺ *Small farmers should not dismiss buying a larger used tractor if it will fit their operation and if the price fits their budget.*

GOOD HEALTH

I am talking about myself, so let me pursue a tangent for a moment and make you aware of the good physical wellness you need to farm, whether or not you use machinery. Many people come into farming after years of sitting at a desk and getting little exercise. This is especially true of people just moving to the country as retirees, but may be valid even for young people who have simply tired of city life. (A recent study of the *Small Farm Today* readership found that 47 percent had been farming less than 5 years.) The equipment recommendations I make are for younger, reasonably physically fit adults. If you are not younger or physically fit, don't give up, though. Go on a diet, work into things slowly, and use equipment to enhance the physical strength you do have.

Remember, "Where there's a will, there's a way." I farmed full time for 30 years, and even with the challenges I mentioned above, I still farm 80 acres, and still do 80 percent of my own farming chores. (I hire out the rest.) I have sheep, hogs, and poultry for livestock, raise open-pollinated corn, and grow a market garden. The point is, if you want to do it, go for it — just use your common sense and, if need be, hire neighbors or young people for the really heavy work. I know an elderly retired farmer who farms stock tanks full of soil about waist high. He makes them accessible to his golf cart so he can travel about and tend the plants himself. This is healthy for him and healthy for his community. You can do it if you want to.

Acquiring Machinery

The first decision to make about machinery is the level of use for which you need it. People who are growing their own food on a small scale, for instance, can get by with very little equipment — there are many ways to prepare the ground with hand tools for planting, rather than by using a tractor or a big rototiller. Mulching the ground the year before it is needed (softening the soil, killing the vegetation) and then opening furrows in the mulch was a technique practiced by the writer Ruth Stout in her no-work gardens, and she successfully produced large harvests of vegetables. Those who advocate raised-bed gardening may use a spade or a special tine tool to loosen the beds. Before you can make choices, you must know what your options are. Let's discuss the varieties of machinery available, starting with the original supplier of horsepower.

Draft Animals

Draft animals were the original "machinery" used on farms to save time and effort. Although most farmers have replaced them with tractors and combines, they are still a viable tool today. Sustainable agriculture is very broad in its definition, and there is more than one model for farming success. The Amish, for example, are a successful agrarian society who use draft animals. Draft horses and other draft animals can be very useful on a small farm and may be more economical for some jobs than a tractor. The primary argument for tractors is that they do the work faster and at less cost. A tractor can plow 12 acres in a day, while four horses would take 3 days. But if you only have 2 acres of fields, two horses can plow the entire plot in a single day, and a tractor will not save that much time.

Costs for draft horses have remained relatively fixed as time goes by — feed costs have increased some, but not exponentially; horse-drawn equipment is rare, but relatively cheap (although it is

Draft animals may provide an economic power source for your small farm if you are willing to take the time to learn how to use them from an experienced teamster.

increasing in cost as more farmers shift to draft animals); and horses are much cheaper than new tractors and similar machinery. Horses also reproduce themselves, ensuring a continuation of available equipment.

However, if you decide on draft animals as your power source, there is a definite learning curve. Here are some tips to get you started:

- ❖ Read all you can about the subject.
- ❖ Enlist the help of an experienced horseman or oxen handler (drover) to help you pick out your animals and equipment.
- ❖ Get older, experienced animals until you develop your skills.

It is not easy to drive a 2,000-pound horse with 12-inch dinner-plate hoofs in a straight line 40 inches wide between two rows of corn — and plowing is even harder on you and the horse than is cultivating.

Draft animals must be approached just as other methods of power are — by evaluating cost of the power source (draft animals) and cost of the equipment. When choosing between a tractor and a draft animal, considerations include size of operation, ease of use, whether the animals will be bred and sold, and relative costs. For 200 acres of crops, I recommend a tractor. For 2 to 80 acres, draft animals are an option. There is one other thing to consider: Do you enjoy being around draft animals, working with them, and raising them? If not, a tractor is better for you. If you do, some of the other considerations become less important, and you can find a way to make it work.

Tractors

A tractor is considered the usual indispensable tool for farms today. Tractors are used for tilling, planting, harvesting, hauling manure and supplies, plowing snow, and leveling roads. They

The Amish

The Amish society is more than 300 years old, according to Donald B. Kraybill in his book *The Riddle of the Amish Culture.* He notes that from a small group of 5,000 in 1900, they have blossomed to more than 100,000 in North America today. The question Kraybill asks is, How do they manage to flourish in the midst of industrialization?

The typical Amish farm family has one or two driving horses and six to eight horses or mules for fieldwork. Families who no longer live on the farm have one or two horses for transportation. Amish are not required to own a horse but it is the assumed mode of transportation, as cars are forbidden.

The horse, according to Kraybill, represents key values involving tradition, time, nature, and sacrifice. Horses make for a slower pace that conforms to nature: Horses don't have headlights, for example, so they can't be used in the fields at night. Daily travel is limited to about 25 miles, and the size of Amish farms is typically about 50 acres. The horse also links the Amish with the natural rhythm of the seasons. And the horse keeps the Amish out of cities.

The Amish succeed by having low input costs, minimal needs, and a strong support network of family, friends, and neighbors. They produce quality goods in traditional ways, and avoid many of the costs of an industrial society. They maintain high soil fertility by mostly natural means, and often utilize the crops themselves or use direct markets. They believe in planning crops to have work throughout the year. The Amish have proved that horses can be cost-competitive with modern machinery, but it takes knowledge, understanding, and a love of animals to succeed.

replaced draft animals on most farms, and are the obvious choice for those without the time or temperament to deal with living creatures as tools. Tractors range from little more than lawn mowers, to garden tractors (up to about 20 horsepower), to behemoths of more than 100 horsepower. There are also specialty tractors, with narrow bodies or high clearance for use in orchard or vegetable beds.

The key to choosing a tractor for your property is to decide what you can afford for the time saved, what you will be doing with it, and what implements will be attached. On most small farms, there will be only one tractor, so it must be suited for everything from plowing to hauling a wagonload of firewood.

What Will You Do with It?

Consider the size of your fields and the turning radius of the tractor. If you have 5 acres, a 60-horsepower tractor is unnecessary. If you have 40 acres of vegetables, you will probably want a fairly substantial tractor. Most small farms could use a 30- to 50-horsepower tractor with a three-point hitch. Don't overlook a high-horsepower used tractor, though, if the price is right. Jobs come up for which you would be glad to have the extra power. For safety's sake, get a wide-front-end tractor. They are much more stable and safer for all types of farm jobs.

Implements

Tractor implements are rated by category. Category I (20 – 45 horsepower) are usually best for small operations, although Category II (50–100 horsepower) are sometimes appropriate. Those in Category II are usually too expensive and too unwieldy. Tillage tools include moldboard plows, chisel plows, disks, harrows, and mulchers. Planting implements include drills, corn planters, and broadcast seeders. Cultivation tools include cultivators and rotary hoes. Flame weeders, blades, loaders, rakes, balers, mowers — such as brush hogs — sleds, bale carriers, manure spreaders, and trailers are other specialized implements. There are also pull-type combines and corn pickers. If you buy a tractor, make sure you get the best use from it by adding appropriate implements.

Hand Tools

Before you consider a tractor (or a draft animal), look at less expensive alternatives. The first option for farming is hand tools, including hoes, shovels and spades, trowels, transplanters, seeders, drills, rakes, forks, scythes and sickles, shears, calf hooks, field knives, wheel hoes, wheel cultivators and blades, flame weeders, spreaders, wheelbarrows, and wagons. Steve Salt, a vegetable farmer and frequent contributor to *Small Farm Today* magazine, praises hand tools for their low cost, durability, minimal environmental

Implements and Performance

Operation	Implement	Power	Acres/10-hr Day
Plowing	Walking plows		
	8-in.	1 horse	1.0
	14-in.	2 horses	2.0
	Moldboard plows		
	2 bottom 14-in.	15-hp tractor	8.0
	3 bottom 16-in.	50-hp tractor	20.0
	Disk plow 50 in.	20-hp tractor	13.0
Disking	Single disks		
	8-ft onceover	4 horses	15.0
	20-ft onceover	20-hp tractor	60.0
	Tandem disk	70-hp tractor	60.0
Harrowing	Spike tooth		
	10-ft onceover	2 horses	15.0
	32-ft onceover	20-hp tractor	90.0
	20 ft	50-hp tractor	130.0
Cultivating corn/cotton	½ row walking (2 times per row)	1 horse	2.5
	1 row riding (1 time per row)	2 horses	7.0
Cutting corn	Binder 1 row (not shocked)	3 horses	6.5
	By hand (shocked)	Hand	1.2
Picking corn	1 row	20-hp tractor	10.0
	By hand	Hand	1.5
Planting corn	2 rows, 36- or 40-in.	20-hp tractor	20.0
	2 rows, 36- or 40-in.	2 horses	10.0
Mowing hay	Mower 5-ft	2 horses	8.0
	Mower 7-ft	15-hp tractor	20.0
Drilling grain	Dish drill 7-ft	3 horses	12.0
	Disk drill 10-ft	20-hp tractor	25.0
Combine pull-type	Combine 5 ft	15-hp tractor	11.0

From Sherman E. Johnson et al., *Managing a Farm* (New York: D. Van Nostrand Co. Inc., 1946), 173.

impact, maneuverability, and precision. They give farmers a closer perspective on their crops and soil, too, he says.

Examine hand tools in terms of the amount of soil they will work and how much time it will take. A wheel hoe may break the soil faster than will a hand-held hoe, but it may not work in some tough, heavy clay soils. Choose tools carefully. Some hoes are sturdier, and a wide variety of them have specialized functions. A sharp-bladed heavy field hoe allows you to clean a field of weeds much quicker and for less effort than does a dull sheet-metal garden hoe. If

TRUE STORY

Steve Salt lives near Yarrow in northern Missouri. Steve, his wife, and children own 147 acres of land and produce vegetables, herbs, and small fruits on about 7 acres and also produce four or five acres of sweet sorghum. Steve trades labor on a neighbor's crop for the use of the neighbor's processing equipment and boiling pans for the sorghum. The neighbor and he both sell the sorghum when it's ready.

Steve is a diversified market gardener, using mostly hand tools (and some small powered tools) to manage produce for a myriad of niche markets. His main crops are sweet sorghum and ethnic and heirloom vegetables. Steve's sorghum is picked by hand or cut with cane knives by family and crews of neighbors, who receive sorghum syrup in trade. He grows about 300 varieties of vegetables and about 150 species of plants. He sells most of his crops at farmers' markets. Sixty different kinds of tomatoes and sixty kinds of peppers are popular with his Hispanic customers, as well as Mexican herbs and plain old head cabbage. Chinese greens are favorites with his Asian customers. His small fruits are enjoyed by everyone.

An interesting sideline for Steve is that he boards about eighty head of rodeo stock (horses) on the rest of his farm. He feeds and cares for them and calls the owners for any vet care or illness problems. He is also finishing a two-volume book on ethnic vegetables.

you only have a small garden though, you may not wish to invest $40 in a quality field hoe. Most of the time, you will want quality and durability in your tools, so examine them carefully before you buy. Look in the resource listings in the appendix for suggestions.

Small Motors

The next step up from hand tools is a two-wheeled tractor and a walk-behind tiller. These hand-guided machines usually range from 5 to 20 horsepower. They start at about $250 for tilling; most are $600 to $800. At higher prices, they allow power take-off-driven additions such as snowblowers and mowers. These are usually an excellent investment for a small-scale market gardener.

Processing Equipment

Processing equipment requires careful consideration before you invest in it. Most added-value processing will involve large stock, poultry and rabbits, or crops. For large stock — beef, hogs, sheep, goats, bison, elk — processing must be done in a county-inspected or USDA- inspected plant if you plan to sell the meat to the general public. Processing at home is not possible. Crop processing (jams, jellies, bread, for example) involves establishing a USDA-approved kitchen, as discussed in chapter 7.

Processing of poultry — chickens, turkeys, ducks, emus — as well as game birds and rabbits, can be done at home. The amounts allowed and sale conditions vary from state to state, but usually several thousand animals may be slaughtered — call your county and state health departments for more information. For poultry butchering, you will usually need:

* a killing cone to hold the chicken down while you cut its throat to prevent wing flapping from bruising meat;
* a rack on which to hang killed birds to drain blood;
* a stainless-steel table on which to eviscerate birds;
* a scalder to loosen feathers;
* a plucker to remove feathers;

- ❖ tanks full of ice water to cool off birds; and
- ❖ a cooler in which to store birds.

Most home operations stop here; customers will pick up the birds in coolers of ice to take them home for freezing or eating fresh. If you start selling to stores, you will probably need a wrapping machine and appropriate containers. Rabbits and fish will require some, but not all, of the equipment listed for poultry. Some suppliers for this home equipment are listed in the back of this book.

Processing may or may not increase your profits. If you add value to your product, but you increase your labor costs 10 times, or your container costs increase fivefold, you may not actually increase profits. The cost of buying processing equipment or establishing a kitchen will also reduce your profits for a time. The question is whether you will process enough units of volume to increase your profits. Establish break-even costs before investing in the equipment. Will you be able to sell the processed product at a price that will cover your processing costs, or does that put you out of the competitive market? My best suggestion is to start by helping someone do his processing. Decide the best way to do it, then buy your equipment.

Fencing

You may not think of fencing as machinery, but netting and fencing in the right places will certainly ease your burden of labor. Fencing requires careful thought about placement, along with quality materials, correctly used. Poorly constructed fences call for constant repair and livestock roundups. Good fences will pay for themselves in time saved in protecting, managing, and moving livestock, and in allowing the best use of every acre of the farm. Good fences demonstrate pride of ownership, aid land stewardship, and increase the overall value of your farm.

When doing initial planning, make sure your perimeter fence (on the property lines) is able to hold any type of livestock, from hogs to cattle. You need to keep your livestock in and your adjoining neighbors' livestock out. As the old saying goes, "Good fences make good neighbors."

Inside your property, make use of lanes to ease movement of livestock from one end of the farm to the other and from field to field. Lanes should be big enough to graze and drive equipment through, rather than narrow, eroded dirt trails. Build a fence wherever possible in such a way that you can mow with a tractor on both sides of it. This allows easy access for maintenance and care of the fence. Make use of temporary inside fencing for seasonal pastures and crop rotations. Electrical fencing, especially the new high-tensile wire and fast-pulse chargers, is economical and more valuable as temporary fencing around grain crops or for intensive grazing.

Get the best fencing and materials you can for durability and economy. At most farms, this will usually mean woven wire or three or four strands of barbed wire on the perimeter, and barbed wire fences on the interior. Elk farms require heavy, 8-foot-tall fencing. Horse farms often use wood or vinyl fencing. For small stock, fences built with hog panels are good. Chicken wire may be necessary to contain poultry.

Other Machinery

Additional machinery for specialized uses includes hay balers, hay wrappers, and combines. There are specialized tools as well, ranging from sawmills, to tree shakers, to cotton gins. Most of these are too expensive and too specialized to justify purchase for a small operation. Rent or borrow from a neighbor if you can.

Keep in mind that some types of structures cross the line into tools and machinery. Loading chutes greatly ease the task of transferring livestock. Squeezes, headgates, catch crates, and corrals serve the same purpose. To decrease your labor and costs, examine each type of livestock and crop, and consider what will work best to increase their quality and decrease your time invested.

Building a Barbed Wire Fence

1. Mow or clean the area where the fence is to be built.
2. Set a corner post at each end of the fence.
3. Stretch one barbed wire between these two corner posts to establish and line up your fence line. Leave in place; it helps keep out dogs and coyotes. Staple 2 inches above the ground line.
4. Make a common post-brace assembly at each end of the fence.

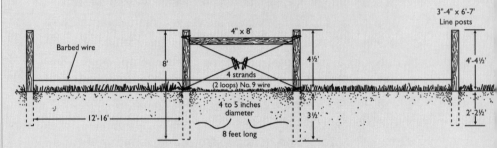

5. Install a line brace each 660 feet on flat terrain, or more if you have hilly terrain to work with.
6. Mark and set or drive line posts 12 to 16 feet apart.
7. Stretch the top barbed wire and staple the wire snugly, but do not drive staples so tight that staples cut into the wire.
8. Stretch three to six lower wires to proper tension and staple.

Materials for a Double-Span End Brace

Material	Diameter	Length
End post	5–6 in.	8 ft
1st brace post	4 in.	8 ft
2nd brace post	3 ½ in.	8 ft
1st compression brace	4 in.	8 ft
2nd compression brace	3 ½ in.	8 ft
4 steel pins	⅜ in.	4 ft
Tension wire	No. 9	100 ft

Note: 1 ¼-in. galvanized pipe may be used for compression braces. Set posts at least 3 ½ ft deep.

Installing Braces

1. Attach braces using the dowel-pin method as shown at right.
   a. Notch the post to fit the brace.
   b. Drill ⅜-inch holes 2 inches deep in post and brace.
   c. Drive a ⅜-inch steel pin in the ends of each brace.

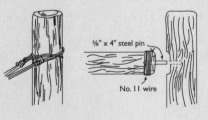

2. Wrap the ends of braces with several strands of No. 11 galvanized wire and twist tight. This adds strength and keeps the brace from splitting.
3. Insert brace pins in holes and assemble the braces.
4. Fasten tension wires.

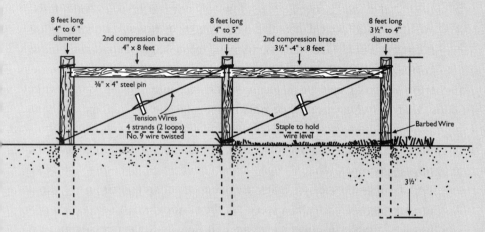

a. Use four strands (2 loops) of No. 9 smooth galvanized wire.
b. Tie wire 4 inches from top and bottom of posts.
c. Staple wire to each post with three 1 ½-inch staples.
d. Twist wire tight with a short stick or board; leave stick in place for later tightening.

Wire Stretching and Stapling

1. Stretch woven wire with a fence stretcher or tractor until tension curve is half its original depth.
2. Use fence clamps (two 2 x 6s bolted together) on all woven wire fences. Attach parallel to and between two wire stays.
3. Attach stretcher chains so there is an equal number of line wires below and above line of pull. One stretcher is sufficient on 26- to 32-inch fences; higher fences require two stretchers.
4. Staple fence to posts on ridges and depressions before the stretcher is released.
5. Cut wire, remove the two vertical or stay wires, then tie and staple to end or corner posts.
6. Use 1-in. staples for hardwood, 1 ½-in. staples for softwood.
7. Drive staples in line posts diagonally with the grain — loose enough to allow wire to slip through staples. Staple wire securely to corner, end, and brace posts.

From Carl Scheneman and Albert Hagan, *Good Fences for Your Farm*, University of Missouri Extension Circular 667, 1956.

Computers

The last piece of machinery I would like to discuss is the newest tool for farms — the computer. I recommend investing in this tool for a variety of reasons. A computer is excellent for storing budgets and income data, figuring taxes, and managing records. There are also farm programs for specific purposes, from keeping track of field yields to managing ostrich operations. E-mail can be a great way to communicate with other farmers and gather information from bulletin boards, such as Sheep-L and Graze-L. Last, with Internet access, you can download all sorts of useful information, from grazing tips, to okra recipes, to site plans.

Computers do not have to be expensive. For most farm uses, you won't need a top-of-the-line model with snazzy graphics. A simple computer that can handle a word processor and an income program will do most of the functions you need. Look at used computers, or low-end models. As always, consider cost and benefits carefully before making a purchase.

Determining Equipment Size

The old adage "Make it yourself, wear it out, use it up, or make it do" is a principle that applies to any small farmer, and to the tools and equipment on his farm. The small farmer's most limiting factors for success are time and capital. Machinery frequently requires large amounts of both. Fixed costs on equipment that is used infrequently is an expense that small farmers do not need and cannot afford if their operations are to be economically viable and thus sustainable. So the question you must ask is not "What do I want?" but, instead, "What do I actually need?"

Farmer Steve Salt offers these guidelines for tool and equipment usage. Human-powered tools — hand hoes, wheel hoes, push plows — are inexpensive and reliable for use on up to about a half acre of crops. Five- to 20-horsepower two-wheel walk-behind tractors can handle from ½ to 2 acres. These could handle even 10 acres, if you custom-hire the heavy jobs like plowing, or you could rent a tractor and plow for the time you need it. If you go beyond the ½- to 10-acre market-garden range and get into 10 to 80 acres or more, a combination of a two-wheel walk-behind tractor and a four-wheel ride-on tractor can be justified, especially on livestock and crop farms. You simply cannot move a 1,500-pound round bale with a two-wheel tractor, nor can you plow, cultivate, and harvest 40 acres of field crops in a timely and profitable manner.

On my own 80-acre farm, I use a combination of one 8-horsepower BCS tiller and one 23.5-horsepower Ford tractor, and have a neighbor haul my share of big round bales, which he bales. Another neighbor picks my corn with his harvester, although I hand-pick some of it early in the season for my hogs.

Variables to Consider

Before pondering costs and ownership, carefully consider these issues:

❖ **Rental or leasing.** Is this possible when we need the machinery, or is there a high demand at that time? How far do you have to move the machinery to get to your farm? Is delivery included in rental?

❖ **Neighbors.** Can you exchange machines with a neighbor, or barter for the use of what you need? This could be as simple as, "I will bale your hay if you combine my corn." Transportation costs must be figured if you are trading with someone other than a neighbor whose land adjoins yours.

❖ **Hiring.** Is there someone in the area who has the machinery needed whom you can hire? My neighbor bales my hay in trade for a percentage of the hay crop. If you hire people you are unfamiliar with, get references — are they reliable and available for your crop?

❖ **Business potential.** If you do not have the acres necessary to justify the cost of the machine, but there is a need for it in your area, can you buy it to do custom work for your neighbors (and yourself)? Some farmers purchase machinery together, but to avoid problems, I recommend avoiding this unless it is done as a cooperative, with all arrangements among parties worked out on paper in advance.

❖ **Used equipment.** Compare new prices and used prices for the same piece of machinery. Is new necessary? If the machinery you need is a low-usage item, used equipment might be your best buy. Many farmers today are using tractors 30 and 40 years old, and some tools that are even older — perhaps adapted from horse-drawn implements. You might question how tractors can even last 20 years. They can because they are tough and well made. If they are well cared for and maintained properly, your machinery should last your lifetime and more. If you buy used, examine the equipment carefully, and, if possible, operate it to get a feel for it. For a tractor, consider its horsepower and the types of equipment you will be hooking up to it.

❖ **Scale of use.** This is really a part of your thinking process about what crops to grow. Look at machine costs as a percentage of the total costs and net profit, according to the crop grown. If you only have one cherry tree and 5 acres of tomatoes, consider machinery for your tomatoes before you worry about the cherry tree.

❖ **Time.** Don't forget the importance of time invested. How does it affect the price you get for your crop?

Economics of Machinery

> **PRINCIPLE:** *Buy the machinery you need at the best price possible, and make it last by faithful maintenance.*

There is a simple key to buying machinery: Buy only what you absolutely need at the best price possible, and make it last by cleaning, adjusting, and using it properly — and storing it away from the effects of weather. If you do not know how, acquire an operator's manual or consult other farmers who are using the same type of equipment you are interested in.

Machinery cost is a fixed cost of production, and according to *Used Farm Equipment,* Northeast Regional Agricultural Engineering Services at Cornell University, its cost makes it important to "minimize machinery costs for a level of productivity and efficiency that will maximize profits."

Machinery Costs

Machinery costs are of two types: ownership and operating costs. You may own a combine, but perhaps you use it three months of the year, at harvest, because it is no good for any other jobs on the farm. A tractor, in comparison, may be used every day, especially if you have a diversified livestock and crop farm.

Ownership Costs

Ownership costs (also called fixed costs), which are about the same amount every year, include the following:

❖ **Depreciation.** The machinery decreases in value no matter how much or little you use it. (This amount also needs to be set aside each year to buy new machinery when the one you have wears out.)

❖ **Taxes.** Fees imposed by the government include property tax (personal and real estate), sales tax, income tax, Social Security, and possibly others.

❖ **Insurance on facilities and equipment.** Insurance is a form of risk management. The risks of fire, theft, flood, and so on could put you out of business if you're not insured.

❖ **Interest,** if you take out a loan to buy the machinery. Interest costs must be evaluated carefully before purchase — does the benefit of the equipment justify the added expense?

❖ **Storage facilities.** If do not already have a shed to store your machinery, you may have to build one. Caring for your tools by keeping them out of the weather is one of the best ways to save on future costs.

Note: These ownership costs will accrue even if your equipment never leaves the machine shed!

TIPS FOR STORING EQUIPMENT

Here are some tips from Harold Tucker, the lubricants technical director at Phillips 66 Company, a division of Phillips Petroleum (Bartlesville, OK), on the proper methods of winterizing machinery.

Postseason Shut-Down

❖ Clean the engine compartment and the outside of the machinery using high-pressure washing equipment, if necessary. When mixed with oil, corrosive material in dirt, such as acids and ash from fertilizer and pesticides, can damage an engine.

❖ Remove residual crops from all farm equipment. Stray pieces of corn or wheat, for example, contain moisture that can corrode metal components.

❖ Overhaul, change oil, and lubricate all machinery as soon as possible after the harvest. Drain the oil to remove contaminants, then refill with new oil for long-term storage. Change all filters.

❖ Observe used oil carefully. Note unusual colors or thickness (viscosity). This could mean a cross-contamination of fluids.

❖ Grease press wheels and clutch parts on planters and fittings on tractors and combines.

❖ Remove rust, and repaint equipment. Keep machinery rust-free.

❖ Remove batteries, clean contact points with baking soda, and reinstall the batteries. Then apply a coating of grease on the terminals to prevent corrosion.

The Northeast Regional Agricultural Engineering Service predicts ownership costs of a new tractor at about 13 percent of its cost and 16 percent of the cost for used tractors, based on 8 percent interest. Ownership cost increases one-half of 1 percent for every 1 percent increase in the interest rate. Other machinery such as a plow or a planter has an ownership cost of about 14 percent for new equipment and 17 percent for used equipment.

Operating Costs

The other machinery cost is operating costs. Operating costs — also known as variable costs — are the cost of using a piece of machinery. Variable costs include fuel, oil, lubrication, filters,

- ❖ Run each piece of equipment at low speed for up to 30 minutes periodically throughout the winter to lubricate gears and internal components, unless the equipment has been thoroughly winterized.
- ❖ Use quality lubricants from a reputable supplier.

Preseason Start-Up

- ❖ Check hoses and connections for leaks before operating equipment. The most common areas for leaks include oil pan seals, transmission connections, and the rear and front main seals.
- ❖ Visually inspect engine oil paths by looking for wet spots on the hoses. Do not run your hands along the hoses during inspection. Engine fluids are under high pressure and a leak may force oil under the skin. With the eyes alone, look at the bottoms of fittings and hoses where fluid will accumulate and drip. In addition, machinery that has been sitting in the same place for a long time can leave easy-to-identify oil spots on the ground.
- ❖ Drain winter oil and change filters. Change the transmission fluid and antifreeze, if needed.
- ❖ Invest in a scheduled equipment preventive maintenance program. Set regular cycles to change oil and filters. During the growing season, preventive maintenance is one element of farming that operators can control.
- ❖ Use oil and hydraulic fluid analyses to help prolong equipment life through early detection of abnormal wear.

repairs, and maintenance. For hand tools, this might be as simple as repairing a handle or sharpening the tool. Sharpening qualifies as maintenance because it makes the tool work efficiently and it lengthens the tool's useful life. Fuel, either gas or diesel, is a cost every time you turn the key and use the tractor. The same can be said for oil, filters, grease, belts, and other machine parts.

When considering the economics of machinery, consider the costs in various ways. First, the actual purchase price: How will you pay for it? (Will you need a loan?) Then consider the fixed and variable costs. Use them to figure your costs per hour and costs per

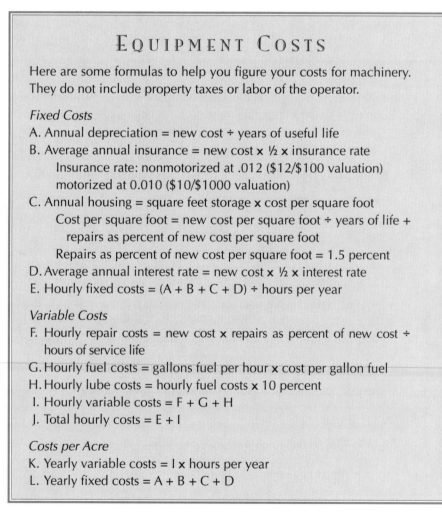

Equipment Costs

Here are some formulas to help you figure your costs for machinery. They do not include property taxes or labor of the operator.

Fixed Costs

A. Annual depreciation = new cost ÷ years of useful life

B. Average annual insurance = new cost x ½ x insurance rate
   Insurance rate: nonmotorized at .012 ($12/$100 valuation)
   motorized at 0.010 ($10/$1000 valuation)

C. Annual housing = square feet storage x cost per square foot
   Cost per square foot = new cost per square foot ÷ years of life +
      repairs as percent of new cost per square foot
   Repairs as percent of new cost per square foot = 1.5 percent

D. Average annual interest rate = new cost x ½ x interest rate

E. Hourly fixed costs = (A + B + C + D) ÷ hours per year

Variable Costs

F. Hourly repair costs = new cost x repairs as percent of new cost ÷
   hours of service life

G. Hourly fuel costs = gallons fuel per hour x cost per gallon fuel

H. Hourly lube costs = hourly fuel costs x 10 percent

I. Hourly variable costs = F + G + H

J. Total hourly costs = E + I

Costs per Acre

K. Yearly variable costs = I x hours per year

L. Yearly fixed costs = A + B + C + D

acre. The fewer the acres, the higher will be the cost per acre. Smaller acreage usually means adjusting to less expensive tools to maximize profitability.

Using the figures from the Equipment Costs chart for a $13,000 tractor used for 600 hours per year, the cost per acre on an 80-acre farm is $60.68. This is before taxes ($120) and labor ($10 per hour x 600 hours = $6,000) are added in; these increase the final figure to $137.18 per acre! This is obviously very high. However, if you had a down payment or a trade-in, this figure might realistically drop to $50 to $60 per acre. Also, most tractors

---

M. Costs per acre = (K + L) ÷ acres used

In the following example, a tractor costs $13,000 new, with a life expectancy of 6,000 hours and 600 hours annual use on an 80-acre farm. It requires 100 square feet of storage space. The storage space costs $3 per square foot new and is depreciated over 30 years. The annual interest rate is 12 percent. Repairs are 90 percent of new cost. Fuel consumption is 2.31 gallons/hour and costs $1 gallon.

A. Annual depreciation = 13,000 ÷ 10 = $1,300.00
B. Average annual insurance = (13,000 ÷ 2) x 0.01 = $65.00
C. Annual housing = 100 x 0.145 = $14.50
 Cost per square foot = (3 ÷ 30) + (3 x 0.015)
D. Average annual interest = (13,000 ÷ 2) x 0.12 = $780.00
E. Hourly fixed costs = (1,300.00 + 65.00 + 14.50 + 780.00) ÷ 600 = $3.60
F. Repairs = 13,000 x (0.90 ÷ 6000) = $1.95
G. Fuel = 2.31 x 1.00 = $2.31
H. Lube = 2.31 x 0.10 = $0.23
I. Hourly variable costs = 1.95 + 2.31 + 0.23 = $4.49
J. Total cost hour = 3.60 + 4.49 = $8.09
K. Yearly variable costs = 4.49 x 600 = $2,694.00
L. Yearly fixed costs = A + B + C + D = $2,160.00
M. Costs per acre = (2,694 + 2,160) ÷ 80 = $60.68

Adapted from Charles DeCourley and Kevin Moore, EC 959: *Selected Fruit and Vegetable Planning Budgets* (University of Missouri — Columbia, Department of Agricultural Economics, 1987).

will run way beyond the 10-year life expectancy most people use for their figuring. My first tractor was a 1948 8N Ford with a 7-foot sicklebar mower, purchased for $800 in 1965. It was 17 years old when I bought it, and I used it until 1975, at which time I sold it for $1,000, without the mower. It is still being used by its third owner.

Another good way to compare costs is to get a custom rate sheet from your local extension office. For instance, in Missouri we

WHAT KIND OF EQUIPMENT DO I NEED?

Depending on the size of your operation, you may need only some of this equipment. There may also be specialized pieces necessary for particular crops or livestock. These lists are merely to give you an idea of what you'll need.

Truck Farm (Small-Scale Vegetable Farm)

Hoe	Rear-tine tiller	Brush hog
Shovel	Greenhouse	Blade
Pitchfork	Small tractor	Disk
Wheelbarrow	Cultivator	Plow
High- or low- wheel hoe	Spray rig	Planter
	Two- or four- wheel wagons	Harvesting equipment

Crop and Livestock Farm

Pickup	Combine	Grinder for feed
Tractor	Corn picker	Feed-storage
Plow	Brush hog	facility
Disk	Blade	Equipment-
Harrow	Grain truck	storage facility
Cultivator	Fences	
Planter	Livestock shelter	

Hay Farm

Tractor	Rake
Crimper mower	Baler
Mower	Wagons

have a custom rate for moldboard plowing ranging from $7 to $15 per acre. Disking runs from $4 to $10 per acre. Cultivating is $3.50 to $7 per acre. Rotary hoe rates are $3 to $5.50 per acre. Corn planting is $5 to $12 per acre. Picking corn in the ear is $17 to $25 per acre. Once you have the custom rate sheet and can compare costs, it is relatively easy to make an economic decision based just on cost.

Of course, more than just cost is involved if you rent or get your farming done on a custom basis. How timely are the custom operations in your area? Can you get references? Some co-ops have small windows of opportunity for harvests such as sweet corn. Weather delays and breakdowns can make ownership seem cheap in comparison to losing your crop. Finally, like everything else in this book, it is up to you to weigh the pros and cons and make the best decision you can live with, labor-wise and money-wise.

Buying Equipment

To buy equipment, you must first know where to find it. Tractor dealerships usually have both new and used equipment and may or may not have some type of warranty period. The second place to look is in your local newspapers or farm magazines for equipment for sale by owner. Auctions are another good place to look, although the enthusiasm of competitive bidders may drive the price beyond its real worth. Know ballpark values before you go to an auction.

Buying used machinery is one way to keep your costs down, though you must be sure you are making a wise investment. First, consider the equipment's age and usage. A tractor's useful life is about 12,000 hours. If you use a tractor 600 hours per year, the useful life is 20 years. Wendell Bower, in his book *Modern Concepts of Farm Machinery Management,* says that the first year of life for a new tractor is the most expensive; each year, operational costs get less until wear and tear causes repair costs to increase. At this time,

Bower suggests trading tractors, because your average use-per-hour costs will increase. This normally occurs at about 10 years of age, or 6,000 to 7,000 hours of use.

Next, figure the costs per hour and the costs per acre to see if this is a good investment. You should be able to figure these out quickly with a pocket calculator. It's not a bargain if it costs too much to use!

Here are some tips on purchasing:

- ❖ Do not get hung up on a particular brand. Do consider, however, local availability of parts, or you may have a long wait while hard-to-find parts are shipped to your farm.
- ❖ Talk to fellow farmers to find reputable dealers and auctioneers with a good reputation. When buying, investigate why the equipment is being sold, and see if there are any maintenance records.
- ❖ Check out the equipment carefully for rust and signs of major accidents. Look for added welds or fresh paint that might disguise these problems. On tractors, examine the tires for wear. Inspect engines carefully, and examine filter condition. You may be able to have analysis of oils and lubricants done at a local lab — check with your extension office.
- ❖ Consider comfort and safety. How comfortable are the seat and shocks? How easy are the controls to reach and operate? On a tractor, does it have a rollover bar?
- ❖ If possible, try out the implement or drive a tractor around before buying it. Make sure everything is working.

AUTHOR'S NOTE

The following tables are outdated, inaccurate, and price varies from one region to another. So why include them? It is extremely difficult to find comparison data for different equipment. These tables will help you begin to compare the costs, usage rates, and lifespan of equipment you might consider for your farm. I have always found these tables useful; I believe you will, too.

Estimated Machinery Costs

Machine	New Cost	Service Life (hours)	Annual Use (hours)	Annual Fixed Cost	Fixed Costs Hour	Repairs % New Cost	Variable Costs Hour	Total Costs Hour	Machine Hours/Acre	Labor Hours/Acre
Plow 1–14" MTD (mounted)	$215	1000	100	$40	$0.40	80	$0.17	$0.57	2.40	3.00
Plow 2–14" MTD	$1,100	1000	100	$188	$1.88	80	$0.88	$2.76	1.20	1.50
Plow 3–14" MTD	$1,600	1000	100	$274	$2.74	80	$1.28	$4.02	0.80	1.00
Tandem disk 6'	$1,250	1200	120	$219	$1.83	65	$0.68	$2.51	0.34	0.43
Tandem disk 8' MTD	$2,800	1200	120	$478	$3.98	65	$1.52	$5.50	0.27	0.34
Rotary tiller 8-hp reartine	$1,150	2000	400	$308	$0.77	65	$0.37	$1.14	4.00	5.00
Rotary tiller 38" MTD	$745	900	90	$127	$1.41	65	$0.54	$1.95	1.30	1.63
Rotary tiller 72" MTD	$1,490	900	90	$251	$2.79	65	$1.08	$3.87	0.65	0.81
Cultimulcher 10'	$3,550	700	70	$618	$8.83	65	$3.30	$12.13	0.21	0.26
Harrow 10'	$266	1200	120	$47	$0.39	65	$0.14	$0.53	0.12	0.15
Plastic layer	$400	1000	100	$81	$0.81	70	$0.28	$1.09	2.00	2.50
Fertilizer spreader spinner	$395	350	70	$122	$1.74	75	$0.85	$2.59	0.12	0.15
Fertilizer spreader 8'	$900	350	70	$264	$3.77	75	$1.93	$5.70	0.42	0.53
Bed shaper	$590	500	50	$112	$2.24	50	$0.59	$2.83	2.00	2.50
Sidedresser 2-row	$860	900	90	$151	$1.68	65	$0.62	$2.30	0.55	0.69
Transplanter 1-row	$475	600	60	$89	$1.48	75	$0.59	$2.07	3.33	4.16
Transplanter 2-row	$965	600	60	$175	$2.92	75	$1.21	$4.13	1.67	2.09
Potato planter 2-row	$3,250	600	60	$554	$9.23	30	$1.63	$10.86	1.67	2.09
Plastic transplanter	$640	600	60	$113	$1.88	75	$0.80	$2.68	3.33	4.16
Planter 1-row	$400	1200	120	$70	$0.58	75	$0.25	$0.83	1.79	1.79
Planter 2-row	$1,050	700	70	$181	$2.59	75	$1.13	$3.72	0.89	0.89

Estimated Machinery Costs (cont'd)

Machine	New Cost	Service Life (hours)	Annual Use (hours)	Annual Fixed Cost	Fixed Costs Hour	Repairs % New Cost	Variable Costs Hour	Total Costs Hour	Machine Hours/ Acre	Labor Hours/ Acre
Multivator 2-row	$2,790	825	165	$742	$4.50	85	$2.82	$7.32	0.75	0.75
Cultivator 1-row	$475	840	120	$106	$0.88	85	$0.48	$1.36	1.38	1.36
Cultivator 2-row	$618	840	120	$140	$1.17	85	$0.63	$1.80	0.69	0.69
Sprayer w/boom or gun	$695	750	150	$192	$1.28	80	$0.74	$2.02	0.21	0.21
Sprayer drop nozzles	$750	750	150	$207	$1.38	80	$0.80	$2.18	0.43	0.43
Air blast sprayer	$4,600	1000	200	$1,252	$6.26	80	$3.68	$9.94	0.63	0.63
Mower 7'	$2,012	1000	100	$340	$3.40	180	$3.62	$7.02	0.66	0.66
Rotary mower 5'	$918	1000	100	$157	$1.57	180	$1.65	$3.22	0.71	0.89
Rotary mower 7'	$1,550	1000	100	$263	$2.63	180	$2.79	$5.42	0.60	0.75
Potato digger 1-row	$3,300	1000	100	$551	$5.51	50	$1.65	$7.16	3.30	4.13
Front-end loader	$2,650	2500	250	$445	$1.78	65	$0.69	$2.47	0.00	0.00
Fork lift 1-ton	$11,500	6000	600	$1,926	$3.21	90	$1.73	$4.94	0.00	0.00
Hole digger 24"	$800	1000	100	$136	$1.36	60	$0.48	$1.84	0.00	0.00
Harvesting aid 2- to 3-row	$3,450	2000	200	$634	$3.17	50	$0.86	$4.03	1.00	1.25
Manure spreader 130 BU	$3,042	1000	100	$527	$5.27	65	$1.98	$7.25	0.00	0.00
Trailer	$950	2000	200	$194	$0.97	60	$0.29	$1.26	0.00	0.00
Truck ½ ton	$8,700	3750	375	$1,526	$4.07	90	$6.50	$10.57	0.00	0.00
Tractor 25-hp	$8,776	6000	600	$1,464	$2.44	90	$3.93	$6.37	0.00	0.00
Tractor 35-hp	$11,791	6000	600	$1,962	$3.27	90	$4.82	$8.09	0.00	0.00
Tractor 40-hp	$13,475	6000	600	$2,238	$3.73	90	$4.56	$8.29	0.00	0.00
Tractor 50-hp	$16,136	6000	600	$2,676	$4.46	90	$5.59	$10.05	0.00	0.00

From Charles D. DeCourley and Kevin C. Moore, EC 959: *Selected Fruit and Vegetable Planning Budgets* (University of Missouri — Columbia, Department of Agricultural Economics, 1987).

Estimated Variable Costs of Equipment (per acre)

| Machine | Variable Costs | | | Labor Costs | | |
	Machine Hours	Tractor	Equipment	House	Cost	Total
Plow	1.20	$5.47	$1.06	1.50	$7.50	$14.03
Disk	0.34	$1.55	$0.23	0.43	$2.15	$3.93
Rototiller 8-hp	4.00	$0.00	$1.48	5.00	$25.00	$26.48
Rototiller MTD	1.30	$5.93	$0.70	1.63	$8.15	$14.78
Harrow	0.12	$0.55	$0.02	0.15	$0.75	$1.32
Fertilizer Sprdr	0.42	$1.92	$0.81	0.53	$2.65	$5.38
Bed shaper	2.00	$9.12	$1.18	2.50	$12.50	$22.80
Transplanter	3.33	$15.18	$1.96	4.16	$20.80	$37.94
Sidedresser	0.55	$2.51	$0.34	0.69	$3.45	$6.30
Planter	1.43	$6.52	$0.36	1.79	$8.95	$15.83
Cultivator	1.10	$5.02	$0.53	1.38	$6.90	$12.45
Sprayer/boom	0.17	$0.78	$0.13	0.21	$1.05	$1.96
Sprayer/nozzle	0.34	$1.55	$0.27	0.43	$2.15	$3.97
Mower	0.71	$3.24	$1.17	0.89	$4.45	$8.86
Potato digger	3.30	$15.05	$5.45	4.13	$20.65	$41.15
Irrigation	0.00	$0.00	$1.65	0.21	$1.05	$2.70

From Charles D. DeCourley and Kevin C. Moore, EC 959: *Selected Fruit and Vegetable Planning Budgets* (University of Missouri — Columbia, Department of Agricultural Economics, 1987).

❖❖❖

FOOD FOR THOUGHT

As you seek out and decide on tools and machinery, remember that the best tool is your own mind. Without it, all of your machinery is just lumps of metal and wood.

❖❖❖

CHAPTER TEN

Farm Management

All of the previous chapters have been about management in one way or another. Now I want you to think of management not as how it applies to any individual part of your operation but as a tool in its own right. Management is what will ultimately provide you with profit — and success.

Successful farm management is two things:

1. Farming is taking all you can from the soil so you can sell the surplus to make a profit.
2. Farming is also putting back into the soil all you can so you can maintain and increase its fertility.

This give and take must be in balance for your farm to be sustainable in the long run — and management is what determines the balance.

≺ *Dairy farms are labor intensive and require astute observation of animal behavior. Grass-based dairies and seasonal dairying reduce labor requirements tremendously while still providing adequate income.*

For instance, I have a small, 900-square-foot hog lot on a very gentle south slope (great for my early garden). There is a portable hog house at one end of the lot. The first time I used the lot, it was solid, tough fescue pasture, where domesticated vegetables would quickly be cornered and disposed of by weeds. The three 40-pound feeder pigs I bought plowed up the fescue free of charge, ate lots of grubs and weeds, and generally made the land ready for the vegetable crop that followed.

I have continued this program because it works well. Each pig furnishes an average of 8 pounds of manure per day, from its 40-pound beginning weight to its 225-pound butchering weight, according to the *Swine Handbook of Housing and Equipment*. Hog manure on a per-ton basis contains 13.8 pounds of nitrogen, 4.6 pounds of phosphorus, and 9 pounds of potassium. Eight pounds of manure per day times three pigs equals 24 pounds of manure a day times a 120-day feeding period equals 2,880 pounds of manure, or about 39 pounds of nitrogen, 13 pounds of phosphorus, and 25 pounds of potassium for my vegetable crop. To help with this fertility program, I sow annual ryegrass and hairy vetch in the lot after the pigs are sold as whole hog sausage for $2.50 per pound. This green-manure crop is then tilled under about two weeks before I want to plant vegetables. Every foot of growth I plow down contributes about 2,000 pounds of organic matter.

Nutrient and usage cycles can be repeated over and over. The hogs could also be run back on cornfields if you wish, or you might substitute poultry or sheep for the livestock portion of this cycle. The only limit is your imagination.

Livestock that is sold off the farm will take with it some of the farm's nutrients, but in my example above, because I purchased the pigs, I also brought in nutrients. If you sell all your hay and do not feed it to livestock on your farm, you are depleting your soil fertility. The best program is to bale your hay and feed it back to sheep and cattle on your hay field and let these animals recycle their nutrients in the form of manure. None of these cycles is ever 100 percent effective, or at least is not practical to be so, but you should strive for 100 percent. If you do not reach this lofty goal, regroup, plan, and go forward again. As they say, "Practice makes perfect."

Recipe for Success

All through this book I have talked about the principles of good farming. These principles are the same whether you live in New Mexico, Missouri, or Alaska. Principles of good farming work anywhere in the world.

Specifics of farming are a different matter — what works for your neighbor may not work for you. Everybody seems to want a recipe for success — Do steps 1, 2, and 3, and Bingo! Success! In farming however, there is no single recipe, because sustainable agriculture is site-specific. In other words, it is what works on your farm with your soil types, your management techniques, your capital, and your monetary requirements. You are the best person to put together the package of principles that works for you. For instance, using farming or tillage practices that conserve soil and improve fertility through the use of cover crops is an important principle. In the north-central sections of the United States, rye and hairy vetch might be your cover crops; in the South, cowpeas might be a better choice. The principle is the same, but the crops change — and, of course, your management of these different crops will be different.

The most important management principle is, "Look at the big picture." How does everything on your farm fit together to make your operation smooth-running and profitable?

Take Stock of Resources

To get a handle on the big picture on your particular farm, you will need to look at specifics. In chapter 1 you asked yourself, "How many acres of tillable ground do I have? How many acres of pasture or timber ground? What farming skills do I have? What equipment do I have?"

In other words, take stock of all your existing resources. In the early 1940s, the Missouri Extension Service had a program called Balanced Farming that expressed just the idea I'm trying to get across here. Unfortunately, "progress" caused the service to drop the program. It was solid farming principle then — and it is still appropriate today. In fact, it is what sustainable farm management is all about. Each resource should be managed to maximize its best use and best profit.

Practice Balanced Farming

Balanced farming is a process of planning and putting together a combination of measures to give optimum (not necessarily maximum) net farm income, to provide conservation and improvement of soil fertility, and other resources. The resources to be combined are land, labor, capital, and management.

The key word is "balanced." One full-time person expends 300 ten-hour days in labor or 3,000 hours per year. If you lay out a farm plan that requires 4,000 hours of labor to accomplish, your labor requirements will be unbalanced if you have only one full-time worker (yourself). When your labor requirements are too high, your production rates (bushels/acre, pigs/litter, and so on) are frequently so low that the additional acres or numbers of livestock won't give you any substantial income growth. On the other hand, if your farm plan requires only 2,000 hours of labor, you will probably get a higher rate of production, because things will be done in a timely manner. The total production possible will be reduced, however, because you will not be fully employed. A combination of crops and livestock keeps you fully employed because the livestock fills the seasonal (wintertime) lull in the crop-production cycles.

If your farming operation is balanced, you will have time for field days, reading research papers, attending workshops, and learning from other resources that can give you the knowledge to improve your operational procedures and your net income at virtually no cost. For example, make sure your labor time is fully employed through the combined use of crops and livestock. Maybe you have discovered through the reading of research papers that side-dressing corn at 12 to 18 inches tall with nitrogen is more efficient, costs less, and is better for plant growth than applying nitrogen at planting time, which risks loosing the nitrogen through leaching before the corn plant is big enough to use it. Knowledge is an on-farm resource supplied by you, the operator.

Optimize Resources

Besides land, labor, capital, and management balancing, balanced farming optimizes the use of your land resources. If you have rolling, steep timber ground, it can be competitive with other crops like corn if it furnishes lumber for building or fence posts and uses the natural inherent fertility of the soils present to do so. Good, level, well-drained soils are optimized by growing field crops like corn, soybeans, vegetables, and herbs. Rolling pasture ground might best be kept in pasture most of the time and rotated with other crops so you might grow 1 year of corn, 2 years of hay, and 2 years of pasture, or it might be in permanent pasture. The balance comes from using the natural inherent soil type already there.

Just because a farm is in timber does not mean it is necessary to bulldoze the trees and plant it in corn, or spray it to death and plant in fescue, like much of the timber ground in Missouri was treated in the 1960s. Timber can produce posts, firewood, wildlife, light grazing, and lumber for building and pens. In my area, taking trees off the steep hills exposes them to erosion, and the hills are difficult to work on with machinery. The best management practice may be to leave the ground in timber — its best use — and, if managed correctly, its best profit. Returns per acre can compare favorably with crops and pasture. If the trees are already there, management is all that is needed to have a resource that is capable of producing a nice income with little start-up capital.

As another management example, I once bought a little creek-bottom farm that had been timbered for logs, but the woodsmen left all the treetops lying on the ground. I could have complained about the expense of clearing it, or tried to work around it, but instead I viewed it as a resource. It made perfect firewood. I cut and sold 100 cords of firewood from those tops at $75 a cord, and only lacked $600 to make the first year's payment of $8,100 on the property.

Knowledge Is Power

> **PRINCIPLE:** *Knowledge is your most important resource.*

Knowledge is your best on-farm resource. What you have in your head and what you can do with your hands is what will direct your management decisions — and no banker or government directive can take it from you. The more you know, the better your decision making will be.

To utilize knowledge, however, you must have information. I cannot emphasize this enough — you must read, read, READ! Attend seminars and workshops, and talk to other farmers, especially those trying new ideas. Profit from the mistakes others have made. There is no point in reinventing the wheel.

Frequent used-book stores (and look in the agriculture, nature, and gardening sections). Check your local library. Look through university libraries for research papers and books from 1910 to 1960. Extension bulletins from this period cover diversified family farms, where cover crops, rotations, and multiple livestock species were once common. Get catalogs from used-book sources advertised in farming magazines. Subscribe to those magazines that cover topics you are interested in. Look for book publishers (such as Storey Books) that put out useful materials. (See the listings in the appendix for more sources.)

The most productive use of any farmer's time is spent in planning, figuring costs, and marketing. You are not lazy when you are planning — you are working smarter, rather than harder. A little modern management applied to time-tested information will help make your farm a success.

Making Choices

You must always make choices. Pastures mowed in June and July remove the majority of early and late weeds before they go to seed. Multiple-species grazing — say cattle, sheep, and goats — is not easy, but will accomplish the same purpose as mowing, minus the fuel and labor costs. Sheep will eat 90 percent of the common weeds, which generally are high in trace minerals important for animal health and reproduction.

Once you have chosen your crops and livestock, marketing decisions must be made (if they were not made first). The ability to sell yourself and your farm is what will provide financial security.

Management Tools

Let me now take a couple of topics from earlier chapters, and explain how they fit into farm management.

Crop Diversity as a Management Tool

Traditional agriculture, as they say, is the only business in the world that buys retail, sells wholesale, and pays the freight both ways. This is not the way to succeed. Today's farmer must be a producer, a marketer, and a salesperson. Part of management is finding a successful balance between production and marketing, and also allotting time for family life, so you do not become bogged down with work, feel discouraged, and quit farming.

Sustainable agriculture and direct marketing are both tools of management. Sustainable small-farm management is based on diversification. Many modern farms are monocultures (all corn, all hogs, all soybeans). They are usually that way because producers find out what crop they can grow best on their land with the least management. Switching this type of thinking to an alternative crop does not improve it. When exotic animals came on the farm scene and farmers started raising emu and elk, for instance, they claimed they were diversified. They were not. When all you raise is one animal, you are not diversified, even if the animal is exotic. When you raise only one animal on your farm, you are extremely vulnerable to price fluctuations and labor extremes.

Monocultures cause other problems, too. Continued use of one crop, especially if it is a heavy feeder (using large amounts of N-P-K) like corn, depletes soil fertility. Just changing crops is not a rotation. A true crop rotation alternates heavy- and light-feeding crops to complement their nutrient requirements and increase soil fertility.

Crop Rotations as a Management Tool

Rotations increase the total yield of crops over the years of the rotation. Rotations improve soil by the use of soil-building legume crops; they also help even out the distribution of both labor and machinery use. For instance, a small-grain crop (wheat, oats) eliminates the need for plowing if it follows a row crop like corn or soybeans. Legume hays or pasture seeding following small grains eliminates plowing and sometimes even disking. Normal rotations of 3 to 5 years usually consist of a row crop, a small-grain crop, and a sod crop. Different parts of the farm may require managing more than one type of rotation because of different soil types or

A Sample Rotation

Vegetables can also be combined with traditional field crops. For instance, you might plant corn, followed by potatoes, followed by a small grain, followed by a hay crop. I plant corn in early May (in Missouri). In early August, I sow winter rye and hairy vetch, so the corn ground is partially exposed for only about 75 days, from May 15 to August 1, to keep down erosion and reduce nutrient leaching. The rye and vetch cover the ground the same year the corn is planted and can furnish hay, seed, grain crop, or green manure plowed under in the second year. I will also have legume grasses and hay available for livestock grazing, which contributes fertility and income to the rotation plan. The rye sops up any nitrogen left from the corn crop and holds it in the rye plant to be used the next year. The hairy vetch, a legume, furnishes nitrogen, which is useful for, say, a tomato crop that follows the corn crop. On or before May 10, I will mow the rye and vetch and plant caged tomatoes in the resulting mulch. The mulch helps control weeds, and the alleopathic tendencies (that is, the ability to prevent germination or growth of another plant) of winter rye also help with weed suppression. The rye and vetch mulch holds soil moisture and slows erosion.

varied terrain. Rotations, according to Nicolas Lampkin in *Organic Farming,* have to maintain soil fertility, organic matter levels, and structure, while ensuring that sufficient nutrients, especially nitrogen, are available and that nutrient losses are minimized. Rotations are the main means of reducing weed, disease, and pest problems by achieving crop diversity both in space and in time.

Although I have mainly discussed traditional crops and livestock in rotations in chapter 4, the principle of rotation works equally well with vegetables and livestock. In general, potatoes, sweet corn, broccoli, and strawberries are big feeders of nitrogen, while beans, peas, pumpkins, and lettuce are low users of nitrogen. Of course, big feeders will have to be followed in the rotation by a legume, while low users can be followed by a non-leguminous crop. (*The Knotts Handbook for Vegetable Growers*, by Oscar Lorenz and Donald Maynard, is the old standby for complete lists of all sorts on the vegetables I talk about here.)

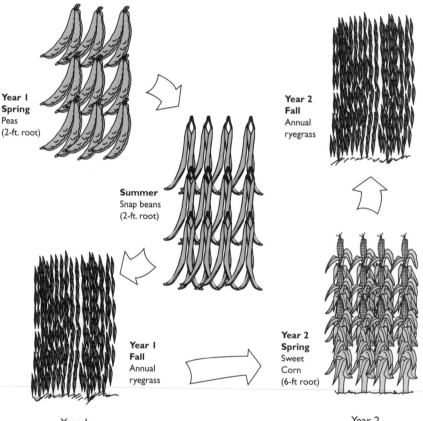

Year 1
Spring
Peas
(2-ft. root)

Summer
Snap beans
(2-ft. root)

Year 1
Fall
Annual
ryegrass

Year 2
Fall
Annual
ryegrass

Year 2
Spring
Sweet
Corn
(6-ft root)

Year 1
(nitrogen-fixing crops)

Year 2
(nitrogen-demanding crops)

In this simple rotation, peas and beans provide nitrogen for the following year's sweet corn, while the ryegrass protects the soil from erosion and provides green manure (organic matter).

Dick Raymond, author of *Joy of Gardening*, has a 2-year rotation that tills under only crops grown on the plot. He uses no leaves, no mulch, no compost, no manure, and no fertilizer. The first year he grows a green-manure crop of peas followed by snap beans followed by annual ryegrass. The second year, he plants a crop of sweet corn (a heavy nitrogen feeder) followed by annual ryegrass.

This rotation alternates nitrogen-fixing crops (peas and beans) with a nitrogen-demanding crop, sweet corn. The annual ryegrass keeps the soil covered most of the time, thereby reducing erosion and nutrient leaching, especially in winter. The winter cover also preserves the earthworm population. The alternating depth of root systems — 6 feet deep for sweet corn and 2 feet deep for peas and beans — brings different soil nutrients into the system. The different root biomasses (ryegrass is dense; peas, beans, and sweet corn are lighter) furnish earthworms and other soil organisms with material to live on. The peas are a spring crop, the beans and corn summer crops, and the annual ryegrass is a fall crop. This spreads out the workload, and the different germination times help with weed control. Last, three of the four crops involved in this rotation can be used as edible or cash crops.

Livestock are needed in a balanced rotation of crops and are important in maintaining soil fertility. Cattle, sheep, and goats in particular can convert low-grade roughage like weeds or cornstalks into salable meat products, while contributing to soil fertility — and they do all the work of spreading the manure. Cattle, sheep, and goats also utilize rough ground that is too erosion-prone to be anything but pasture.

Managing Labor

Livestock spreads your labor into use for the entire year, making you fully employed on your farm. There will be some overlap of crop and livestock chores, but having livestock helps even out the peaks and valleys of labor use. This is just as important for good management as is practicing value-added methods to spread your marketing throughout the year.

For example, on my farm I maintain a Katahdin Hair sheep flock that I lamb on the grass pastures in May and June. Not lambing in cold weather saves me labor and vet bills, while using hair sheep allows me to avoid shearing and other labor normally associated with wool. Although there is some overlap with corn planting, the sheep pretty much do their thing at a time of year when weather and grass conditions are good. I usually buy feeder pigs to be sold as sausage in August or September to utilize early corn and surplus corn. This eliminates the labor of marketing the corn. If I let the pigs graze in part of the cornfield, I eliminate the harvesting of that area. The pigs are slaughtered in December or January, to eliminate feeding during the worst of the winter. Sales of their sausage and ear corn for seed give me some winter income. The lambs are sold for breeding stock or meat from August through April, and their cycle starts over. Basically, this schedule allows me to avoid major labor in the cold winter months, while still getting some winter income through value-added sales. Also, by choosing and breeding for animals that can care for themselves, I eliminate unnecessary labor involved in livestock health care.

Labor is a limiting factor in farm management. No matter how much you love it, you can be efficient at only so much. All your own family's labor should be utilized fully before you think about hiring help, because this labor does not require a cash outlay, as an employee would. In raising vegetables, however, peak harvest labor may require hired help in order to have the best-quality product for sale.

Using Family Labor

If you want your children to love the farm and maybe even go into business with you, the farm has to be a positive experience for them. This means thinking carefully before exposing them to some animals or equipment. Integrating your children into the farm chores, though, will help them to learn responsibility, timeliness, and a work ethic. Unlike many "in-town" responsibilities, the animals must be fed/watered/milked on time, or serious consequences occur.

Teach your children safety in everything they do on the farm whether it is working with livestock, machinery, or around the house and garden. Whenever you give a child a chore, explain not only how to do it but also why he or she is doing it. Emphasize doing a chore correctly, rather than demanding speed.

The Amish do not have central heat. Because the heat is in just one room, the family congregates to that room except at bedtime. This brings the family together, and children learn many things by listening and seeing the interaction and relationship of their parents and siblings. I don't advocate turning off the heat in your house — but allowing full family discussions on the farm at dinnertime, or afterward, in the living room, is certainly beneficial. Children learn by example. You are the example. If you get angry and beat on your livestock, they probably will also. If you drive your machinery fast and do not regularly grease and oil it, they will probably do the same.

Young children, age 5 or 6, can do simple chores like collecting chicken eggs. Children this age need adult supervision at all times, and their size in relationship to the animals' size must be taken into account. An old hen will sit on her eggs to protect them, especially if she is broody. A nest box is at about eyeball level with a young child and the hen can threaten eyes by pecking. One way of protecting your child is to give her a small trash can lid, held up like a shield. The child can then push the hen aside and get the eggs and, at most, will get a harmless peck on the hand. Of course, you need to show the child how to do this, explaining why the hen does what she does, so your child will not only understand that the hen is just protecting her property (the eggs), but will also build confidence to do this.

Children's muscles are developed enough to carry, say, a 5-gallon water bucket that is half full when they are 9 or 10 years old. Do not expect them to work as hard as you do, nor as long. Adult supervision is necessary until they have the confidence and judgment to do the chore on their own. Be sure to praise your children every chance you get. It builds self-esteem and confidence. It makes them want to do more.

Extra care must be taken with larger animals. Small children look like dogs or predators to animals like sheep and, again, the mothers will always want to protect their young. I didn't allow my children around hogs and cattle until they were in their teens. This should also be true of any engine-powered farm equipment. Too many accidents have happened to children on and near tractors and mowers. Wait until they are old enough to understand the dangers.

Make sure to teach older children the full spectrum of farming activities. Take them to the farmers' market and let them run the cash register. Take them to the processing plant to hear the instructions you give. Your children need to learn that successful farming includes marketing as well as growing. Let them have a small plot to develop their own crop or raise an animal, then sell it along with yours. The 4-H program encourages this, although it usually works with more traditional crops and livestock. While I disapprove of some of the overly competitive aspects of some programs, it is an excellent way for your children to interact with other farm families. If your child does his or her research carefully, you may end up with a new profitable crop for your farm.

Hired Labor

Hiring labor for the farm is often difficult. Employ outside help only when you have to, and then get the best available. You must be prepared to pay as much as a job of comparable responsibility in town would pay — unless it is an apprenticeship, where the person is working for experience. When hiring, make sure you see a résumé and interview a person carefully about experience, initiative, and responsibility. Establish a written contract, so both people know what they are getting.

If you must hire outside labor, you'll have to consult a tax accountant, because you may need to withhold taxes and Social Security from their paychecks. You may also have to pay unemployment insurance and worker's compensation insurance. Document employee work with performance reviews. If you need to fire a bad employee, you will need documentation of his infractions to avoid complaints or a lawsuit.

To find hired help, first talk to fellow farmers, who may have suggestions based on whom they have worked with. Older children from nearby families who are looking for some extra cash may be willing to do jobs on a part-time or seasonal basis. Check community bulletin boards at local restaurants or stores for specialized help, such as baling hay. A classified ad in the local paper may also be an option.

Once you have good employees, you want to make sure they'll stay. Your interaction with any employee is important for a good working relationship. Your job is to make the person proud and happy to work for you. Do not treat people as just another expense. Be clear with your directions, and show them what to do. Be fair and honest all the time. Incentives or bonus plans for certain levels of production instill pride of ownership in hired help. Extra money may help on the employees' home front when they have to work late for some reason — an employee's family has to be happy with the job, too, or you will not have an employee for long.

Planning for Farm Efficiency

Farmstead arrangement is very important to efficient labor use, especially for building locations. According to *The Farm Management Handbook,* every 100 feet of unnecessary distance between the house and farm adds up to 14 miles of travel a year for each daily round trip. If you have a barn about 1,000 feet from your house and can move it 500 feet closer, you will save about 700 miles of walking each year.

The size and shape of your fields also affect labor. The smaller the field, the more times you have to turn the tractor when tilling the field, and the less time you spend doing productive labor. The larger the field, however, the less important this savings becomes. Obviously, a garden plot would be better to till by hand or with a two-wheel tractor rather than with your 45-horsepower tractor. You also have to consider leaving enough "waste" space between your plot and the fences to allow for turning your equipment. There is no optimal size or shape, except what will work best on your farm. So when planning your fields and pastures, look at your maps carefully, and plot the best

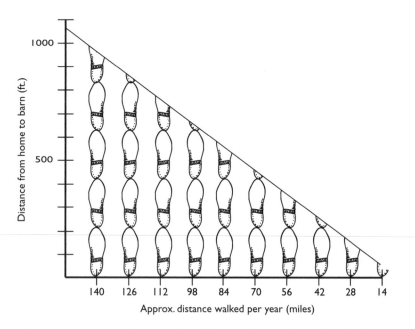

Every step you don't take is time that can be used productively elsewhere on your farm.

placements of access lanes, buildings, pastures, gardens, and fields. Keep walking distances and future uses (rotations, new crops, for example) in mind while planning. No matter how carefully you plan, some new circumstance will change it — but if you have a good plan, adjustments will be less costly in time, labor, and money.

Building Reserves

Reserves are another method of securing financial safety: They will help you through unsteady markets or shaky financial periods. To build reserves requires careful management.

Money

The most obvious reserve to build is that of money. I assume most of you have discovered that money not put in a savings account is usually spent. Setting aside a small amount each month for savings will be a great help. Also, when you buy machinery, it will depreciate each year. *Depreciation* is the cost spread over the life of the equipment. For example, consider a tiller that costs $1,000 and has an expected lifespan of 10 years. The depreciation equals cost divided by expected lifespan ($1,000 ÷ 10), in this case $100 per year. Set aside an amount equal to the depreciation, so you will have most of the cash needed to replace it when it wears out. If your markets are uneven throughout the year, you will need reserves for slower periods. It is always better to skimp a bit in summer than to have to do without in winter.

Having extra monies reserved will also let you take advantage of sales or advertising opportunities without interrupting your present cash flow. It can get you through droughts and floods, and allow vacations and seminar attendance. A small fund to increase your knowledge through book purchases, magazine subscriptions, and conference attendance will help ensure your continued success.

Crops and Livestock

The most obvious resource to reserve other than cash is your crops and livestock. Finding a way to preserve your crops for winter feed will save you money. By saving ear corn, baling hay, and storing potatoes, you are making your crops work through additional seasons. A good hay year should result in extra hay in the barn for a reserve against a bad hay year. Adding value is another way to build reserves, by ensuring that you have products to sell throughout the year.

You can build reserves in livestock by saving back females, which reduces your replacement costs. You save back females by keeping your best young ewe stock each year, while culling the rest for sale. All ram lambs must be sold. Rams should be purchased every year or alternated to prevent inbred genetic problems. This practice will ultimately improve the quality and reproductive performance of the livestock, as you select for the best stock for your farm and management methods. This is another kind of reserve, a reserve of quality.

Quality reserves are also possible with other aspects of your farm. Build your soils with cover crops and careful rotation of crops and livestock — this will result in a reserve of nutrients for future crops. Quality soil is the reserve that forms the basis of farm sustainability.

Reserves can also involve how you plan the use of your farm, with resources both large and small. Timber growing on your farm will always be a good resource for the future, whether in sales or in firewood. Salvaging equipment and structures for other uses is an excellent way to save money. Wire can always be conserved. On my farm, I tie small bits of wire to fencing so I always have a supply on hand. I use it for tying together fences, fixing machinery, tying down water troughs, closing gates, and a multitude of other uses. Lumber and fencing can always be used for repairing similar structures, or save them for future projects.

Make a list of how many different ways you can build reserves on your farm. Examine each crop and type of livestock, your land, your machinery, and structures. Decide what is best to do, and plan how to do it. With careful management, you will weather any storm.

❖❖❖

FOOD FOR THOUGHT

Diversity, rotations, and reserves are all different aspects of farm management, as is the information throughout all of these chapters. Management covers production cost, production expenses, labor, machinery labor and use, land use, soil maintenance, crops, livestock, budgets, diminishing physical output, increasing mental effort, and maximizing economic returns. To sum up farm management is to say that it is all about how you think about the various aspects of farming in relationship to one another. Your management will determine your success.

❖❖❖

CHAPTER ELEVEN

Where We Are Going

My grandmother always used to say, "If you don't know where you're going, how will you know when you get there?" To really know where we are going, first we must know where we have been. The history of agriculture does much to explain how we got to where we are today, and from this, we can determine where we want to be.

Land and Farms

In the early days of agriculture in this country, it was common practice just to move to a different farm when the soil got poor on the land you were farming. Land had little monetary value and was used to encourage people to homestead various areas where railroads and other businesses wished to have customers.

Then dust storms and the Depression hit; Americans became concerned about their food supply and the conservation of our soil. "Land — they're not making any more of it!" was a common quote.

◄ *Direct marketing whole-hog sausage enables you to capture a niche market that is far more profitable than hauling your hogs to the sale barn and saying, "What will you give me for them?"*

With the increasing use of the automobile after World War II, rapid movement became available from cities to the country, and vice versa. People were interested in independence, self-sufficiency, and the good life. A little piece of land and a house outside the city was the American dream. Factories and businesses followed this path to give their employees a taste of country life, and farmland started disappearing. Later, as land prices rose and the best return on money was from developments, ranging from housing to shopping malls to recreation centers, farmland began vanishing, to be replaced by concrete and asphalt. Today, during every single minute of every single day, we lose 2 acres of farmland to development.

Throughout the 1940s and 1950s, a number of writers came forth and exposed various agricultural policies and practices. Around 1943, Louis Bromfield, author of *Malabar Farm*, was warning of future hunger caused by disappearance of farmers. A New York City newspaper headline read, "City Facing First Famine in Our History." Edward Faulkner's *Plowman's Folly* questioned the use of the plow due to its destruction of the soil. Arthur Moore, in *The Farmer and the Rest of Us*, noted that of the 6 million farms in the United States in 1945, one-third of them averaged 20 acres in size and sent about $100 worth of food to the marketplace in a year. They produced about 3 percent of the production. The 3 million top-producing farms grew 90 percent of our farm commodities. I include these comments and figures for two reasons: First, many of the ideas from the 1940s and 1950s have become the sustainable agriculture ideas of the 1990s and the new millennium. Second, we must recognize that farmers and agriculture are disappearing from the American countryside and mind. But there is good news! Though the number of farms has decreased in recent years, the number of small farmers is increasing and will likely continue to do so. In the 1990s, we have about 1.9 million farms; 75 percent of them are small farmers, with 178 acres or fewer, who produce 20 percent of our farm commodities.

Given society's choice of where its food is grown and how it is grown, this could well be more than 50 percent in a few short years. Local farm production leads to stable local communities and citizen participation in local government. Small farms furnish a set of children with a work ethic and a good set of values. Isn't this the kind of nation we want to live in? We need more farmers growing more crops for different reasons, not fewer farmers going to town for more financially successful jobs.

Historical geographer Paul B. Frederic notes that situation is of critical importance in terms of farm characteristics, as relative location determines how the use of a particular site changes through time. Changing proximity to markets leads to change in farm functions. According to *Farming in the Midwest 1840–1900,* edited by James Whitaker, in most cases, farms that retain their pioneer atmosphere are those most remote from population centers. This is still true today, as cattle is ranched in the traditional manner in the ranges of the West, and Appalachian farmers continue their traditional subsistence farming. Population centers tend to support smaller farms with direct-marketed products — the antithesis of the modern "traditional" farm. Small farm agripreneurs need to be located within 40 to 50 miles of a city for the best marketing results.

Industrialization of Agriculture

Originally, farms were small, and mostly fed just the families who lived on them. With the growth of cities, some farms started producing food and fiber for urban dwellers. As time went by, farms grew large and mechanization came on the scene in the form of reapers, binders, corn pickers, and so on, up to the modern-day combine. John Deere developed the first steel plows, and tractors eventually replaced the horse on farms as the main power source.

By the 1950s, agriculture had become fully mechanized. Hybrid corn was the rage, and farmers began to use chemicals to reduce labor costs. Farms became more specialized, and diversification and its benefits gave way to new agricultural thinking. "Fencerow to fencerow" and "Bigger is better" became the watchwords of the young tigers of the 1970s.

The rural crisis of the 1980s brought the agricultural dream crashing down. Out-of-control land prices, combined with a drop in commodity prices, killed a lot of farms. Suicides increased, along with divorces; dreams were destroyed and families were torn apart. The rural crisis brought about a change in agriculture. Farmers began to question the land-grant institutions and they began to question the way they farmed and why. Farmers began to really study farming, and, in particular, marketing. The survivors of the '80s made it either because of excellent equity positions or because they were diversified farms that developed niche markets for traditional or nontraditional crops and livestock, or a combination of both.

The rural crisis was followed by the consolidation of agricultural processing — now, for example, four companies process 87 percent of all the beef in the United States. And other commodities have similar numbers. The same consolidation has happened with the chemical companies, which bought out the seed companies, giving them virtually complete control of major crops like corn, wheat, and beans, from production through processing and marketing. Also, in the 1990s, seed companies began charging a technology fee in addition to the cost of the seed. Recently, the USDA and Delta Land and Pine Company developed a "terminator" gene that makes saving seed impossible, because it is genetically programmed not to germinate the second year. This has received so much bad publicity that Monsanto (which bought out Delta Land and Pine) chose not to proceed with it, but they have continued to develop other varieties of terminator genes. Other companies are also developing terminator technology. The problem with terminator technology (other than the risk of genetic contamination of your open-pollinated crops) is that it forces you to buy from the seed companies every year, rather than being able to save your own seed.

Again, the good news is that even as farm numbers shrink and major companies take over more of the market, the number of small farms is increasing. Small farms of 178 acres or fewer continue to gain in numbers at the rate of 2 percent a year and will continue to do so for at least the next 10 years. I firmly believe that this rate of increase will become even higher as more farmers opt out of traditional marketing and into direct marketing, either by themselves or as part of a cooperative.

Farms and People

To begin with, everyone was a farmer. Many of our early politicians were farmers, and some, like George Washington and Thomas Jefferson, had extensive landholdings in which they experimented with crops and practiced growing legumes for cover crops and soil improvement. As people moved to the cities, the percentage of farmers started to shrink. The average person today is four or five generations removed from the farm, and the people making agriculture decisions have little or no experience with farming.

If we want agriculture to evolve into a more sustainable pattern, the four major groups of importance to agriculture — farmers, lenders, consumers, and institutions — must be convinced at the same time that these changes are beneficial. This will require educating these people as to what farming is about, and why sustainable practices will ultimately benefit everyone — and make them more money.

Success on Sustainable Farms

Despite the somewhat gloomy note struck by my brief history of agriculture, there are more opportunities to start and succeed on a farm today than ever before. Common sense, along with appropriate technology and direct marketing, has put the small-farm dream in reach of anyone who wants to try to achieve it. It can even be done without first going into debt up to your ears to get started.

The whole purpose of this book is to give the principles of good farming, which will work on any type of farm and in any state. The way you picture your farm and its goals and relationships is the Big Picture principle, the balance between your crops and livestock, both locally and globally. The smaller the acreage you have, the more important your level of soil fertility, so that you can produce the necessary volume to be profitable.

Establishing your goals and working to achieve them will create a successful small farm. The most important thing to remember is that the goals must be the entire family's goals. Direct marketing is mandatory for success. To gain the volume needed for success in your monetary goals, you must receive retail prices for most, if not all, of your production. As I write this sentence (November 1998), I will make $100 to $130 per head in net profit on my hogs by selling them as sausage rather than the current market price of 18 cents per pound from the packers. You can do the same.

Farm planning is a road map of how to reach your destination. Farm planning is a principle. Selecting your enterprise correctly to match your farm, your resources, and your management level are important to success. Spend your time analyzing budgets and profit potential.

Machinery and hand tools are labor-saving devices to enable you to do more in less time or to do it in a timely manner. Sweet corn, for instance, has a one- to two-day window of harvest for the picture-perfect ear. Planting every 2 weeks spreads your marketing and marketing risk of timeliness.

Farm management is the process of taking all the principles in all the chapters and making them work together. When you do that, your farming operation will run smoothly, like a well-oiled machine.

I'm sure that as you go along, you will find more principles that are specific to your farm and soil type, but what we have discussed are the basics; with these few, you can start and you can succeed!

First Steps

I have met many people during my years as publisher of *Small Farm Today* magazine who say they want to farm, but it seems they never start. Starting a project like farming, especially if you have no experience, can be quite frightening, but also challenging. You must see yourself as an agripreneur.

Experiment. If you are already farming, but want to move to more-profitable alternatives, set aside a small area to work with and start experimenting. If you are new to farming, start your new farm now, even if it is in the backyard of your subdivision. Start researching, start plotting a business plan, start visiting farmers' markets and talking to fellow farmers. Grow into farming, but do not borrow yourself into farming.

Rent land. There are lots of ways to start — if you don't have any land, rent some. Apprenticing on a working farm is another way to gain experience, with part of your salary being used to save for some land. Use of a vacant lot in the city can often be had merely for clearing it. You can then grow produce on it rent-free.

Learn as you grow. You can grow enough to feed one adult for one year on 1,000 square feet. Do it yourself — it's good practice. Sell some of your produce to your neighbors. Keep records, look at your soil, and examine how it works. You can accumulate tools and equipment, paying cash and storing them until you get your land if you have a plan.

Always watch your bottom line. If you cannot make a profit on 1 acre, having 10 acres or 80 will only make the project 10 to 80 times worse. Thinking equals profits. The more you think about farming and study farming, the more opportunities you will see for reducing your costs, increasing your profits, and becoming sustainable. If you simply reduce your expenses 10 percent, you increase your net profit 40 percent. Food for thought: If your produce or meat is fresher and better than store-bought, why should you sell it for 20 percent less than a store? If you work hard to produce a high-quality, "best-of-the-best" product, you deserve to get paid for it. Price your products at what they are worth.

When should you start farming? If you are talking about traditional agriculture — corn, wheat, and soybeans, or cattle, hogs and sheep — the anwer is to start at the bottom of the price cycle or seasonal low. All commodities prices run in cycles between low supply/high demand/high prices and high supply/low demand/low prices. Crop cycles depend on whether there are any surplus or deficits in previous years. Cattle cycles are normally long, whereas hogs and sheep — multibirth animals with short gestation periods — are normally shorter. It should be noted, however, that environmentally controlled housing, especially for poultry and hogs, has played havoc with seasonal cycles for livestock. Further consolidation of agriculture has essentially caused the marketing system for commodities to fail because there is in effect no supply and demand essential for price discovery. Do not embark on this type of farming (traditional agriculture with conventional markets) unless you want to be a contract grower making no management decisions and shouldering a large portion of the financial risk of farming, which is Land, Labor, and Machinery. However, I would encourage you to start farming using the methods I talked about in this book!

Be a Spokesperson

Much as I would like to see us have 6.5 million farmers again, we have only 1.9 million. Two percent of the population feeds the other 98 percent. To make the changes I have talked about, to keep your farm on an even keel, you must communicate your needs to others. Because the politicians who make the laws and the people who vote on them are four or five generations removed from the farm, it is up to you to educate them. Every man, woman, and child involved in agriculture must be a spokesperson for the agriculture industry. In that role you can reaffirm the farm and food connection, and at the same time build relationships with your customers.

My Vision

Throughout this book I have talked about the contributions that small sustainable farms make to the community and to our society. I also believe we must have a sustainable society to have a sustainable agriculture. Just reading the words seems to give me roots in something permanent — roots in a better place to live and grow.

We easily have room for many more small farms in the United States. This is one of the few bright places in American agriculture because small farms are increasing. In the future, I see small sustainable farms as being the predominant type of localized agriculture. More small farms will be involved in processing their own products, and there will be many regional products like Ozark wines, and Boone County hams, and Washington apples. The regionality will be significant as we shift to a more seasonal diet that goes along with a decentralized agriculture. With more people connected to agriculture, we will have a better understanding of our world, and will understand that what we do today has consequences for our children and their children.

Small farms provide more time for interaction with family and neighbors. The neighborhood will be very broad, as computers and the Internet help small farmers share and gain knowledge with each other around the world. In the future, I believe we will spend much more time thinking, rather than doing, and as a result, we will become more at peace with one another and the world in general.

Everything in life is dependent on our soils, the sun, and water. When larger numbers of people are involved with the soil and the basics of life, food, shelter, and clothing, the world will become a better place, because more people will appreciate and understand how we get these basic tenets and how important they are. In today's world, this is easy to forget, because we are a mobile society that can move away from the consequences of our actions; if the soil is poor, plow up another field, or sell the farm and get another. In the future, we will not have that luxury; we must improve what we have. A small farm must be fertile to provide for the family's lifestyle and livelihood.

Small farms hold the key to reconnect people with the land and the food they must have to sustain life. Small farms are an ongoing process of discovery about life and people. Small farms are the wave of the future, leading to a better world.

It is now your chance to join this movement. You now have all the building blocks — the principles of a successful small farm. All you have to do is start.

Happy and profitable farming!

Note: If you have comments or questions, I invite you to contact me in care of the publisher, Storey Books.

Metric Conversions Chart

Unit	Metric Equivalent
Area	
1 acre	0.405 hectare
1 square foot	0.93 m^2
1 square inch	6.45 cm^2
Length	
1 inch	2.54 cm
1 foot	30.5 cm or 0.305 m
1 mile	1.613 km
Mass	
1 pound	0.454 kg
1 ton	0.907 metric ton
Temperature	
0°F	–17.78°C
32°F	7.6°C
75°F	23.9°C

Resource Lists

In this appendix, I have provided a list of resources that I have found valuable in small farming endeavors. First, though, I wish to promote two resources without which you would not be holding this book:

❖ Storey Books, PO Box 445, Pownal VT 05261; 1-800-441-5700, http://www.storey.com. Storey publishes a wide variety of farming books, including this one. I particularly recommend the *Guide to Raising* farm animals series.

❖ *Small Farm Today* magazine, 3903 W. Ridge Trail Road, Clark, MO 65243; 1-800-633-2535. Bimonthly. I started this how-to magazine of alternative and traditional crops, livestock, and direct marketing in 1984 to provide information to small farmers and small acreage landowners. It led eventually to this book.

Books

There are a multitude of useful books; those listed here are ones I use frequently. The catch is that many of them are out of print. Check with your local bookstore to see if you can order them. And check your local and university libraries, and visit lots of used bookstores — you may come across a treasure.

Alternative Livestock

Benyon, Peter H., and John E. Cooper, et al., *Manual of Exotic Pets* (Ames, IA: Iowa State University Press, 1991).
Guide to exotic pets — mammals, birds, and reptiles.

Birutta, Gale, *A Guide to Raising Llamas* (Pownal, VT: Storey Books, 1997).
In-depth llama guide.

Johnson, Jim, James Harvey Johnson, and Stanley T. Weiner, *Husbandry and Medical Management of Ostriches, Emus, and Rheas* (College Station, TX: Wildlife and Exotic Animal Consultants, 1992).
Short but detailed guide to ratites.

Kyle, Russell, *A Feast in the Wild* (Oxford, UK: Kudu Publishing, 1987).
A look at the potential food use of several exotic species.

von Kerckerinck, Josef, *Deer Farming in North America* (Rhinebeck, NY: Phanter Press, 1987).
Excellent guide to raising fallow deer.

Yerex, David, and Ian Spiers, *Modern Deer Farm Management* (Carterton, New Zealand: Ampersand Publishing, 1987).
Good deer farming guide.

Aquaculture
Brown, E.E., and J.B. Gratzek, *Fish Farming Handbook* (Westport, CT: Avi Publishing, 1980).
Raising food, bait, and tropical fish.

Huner, Jay V., and E. Evan Brown, eds., *Crustacean and Mollusk Aquaculture in the United States* (Westport, CT: Avi Publishing, 1985).
Raising crawfish, shrimp, clams, and more.

————, ed., *Freshwater Crayfish Aquaculture* (Binghamton, NY: Food Products Press, 1998).

Cattle
Morrison, Frank B., *Feeds and Feeding*, 22nd ed. (Morrison, IA: The Morrison Publishing, 1959).
The bible of livestock feed.

Salatin, Joel, *Salad Bar Beef* (Swoope, VA: Polyface Farms, 1995).
How to profit from a small beef cattle operation.

Smith Thomas, Heather, *A Guide to Raising Beef Cattle* (Pownal, VT: Storey Books, 1998).
Handling, breeding and care of beef cattle.

van Loon, Dirk, *The Family Cow* (Pownal, VT: Storey Books, 1976).
The basics of a family cow.

Draft Animals
Miller, Lynn R., *Work Horse Handbook* (Sisters, OR: Farmers Book Service, 1985).
Guide to using draft horses.

Telleen, Maurice, *The Draft Horse Primer* (Waverly, IA: Draft Horse Journal, 1977).
Selection, care, and use of work horses and mules.

Earthworms
Barrett, Thomas J., *Harnessing the Earthworm* (Eagle River, WI: Shields Publications, 1947).
The original book on earthworm culture.

Ernst, David, *The Farmer's Earthworm Handbook* (Brookfield, WI: Lessiter Publications, 1995).
Managing earthworms to improve soils.

Gaddie, Ronald E. Sr., and Donald E. Douglas, *Earthworms for Ecology and Profit*, Vol. 1 (Ontario, CA: Bookworm, 1976).
Earthworm farming.

Lee, K.E., *Earthworms: Their Ecology and Relationships with Soil and Land Use* (Orlando, FL: Academic Press, 1985).
A technical guide to worms and soils.

Minnich, Jerry, *The Earthworm Book* (Emmaus, PA: Rodale, 1977).
How to raise and use earthworms.

Field Crops
Alternative Field Crops Manual (Madison: University of Wisconsin-Extension, Madison/University of Minnesota Center for Alternative Plant and Animal Products, 1992).
A collection of papers on alternative crops, from adzuki beans to wild rice.

Lampkin, Nicolas, *Organic Farming* (Ipswich, UK: Farming Press Books, 1990).
Organic farming in Europe.

Martin, John H., Warren H. Leonard, and David L. Stamp, *Principles of Field Crop Production*, 3rd ed. (New York: Macmillan, 1976).
Basics of crop production.

Robinson, Raoul A., *Return to Resistance* (Davis, CA: agAccess, 1996).
Breeding crops to reduce pesticide dependence.

Specialty and Minor Crops Handbook, 2nd ed. (Oakland: University of California DANR, 1997).
Profiles of 63 specialty and minor crops.

Gamebirds
Mullin, John, *Game Bird Propagation* 5th ed. (Goose Lake, IA: Wildlife Harvest Publications, inc., 1994).
Breeding gamebirds.

Woodard, Allen, Pran Vohra, and Vern Denton, *Game Bird Breeders Handbook* (Blaine, WA: Hancock House, 1993).
Raising pheasant, partridge, and quail.

Gardening
Bartholomew, Mel, *Cash from Square Foot Gardening* (Pownal, VT: Storey Books, 1985).
More information on square foot gardening.

———, *Square Foot Gardening* (Emmaus, PA: Rodale Press, 1981).
Planning a garden in square feet.

Coleman, Elliot, *The New Organic Grower,* 2nd ed. (White River Junction, VT: Chelsea Green, 1995).
Organic tools and techniques for home and market gardeners.

Poincelot, Raymond P., *No-Dig, No-Weed Gardening* (Emmaus, PA: Rodale Press, 1986).
Organic gardening without tilling.

Greenhouses and Solar Gardening
Edey, Anna, *Solviva* (Martha's Vineyard, MA: Trailblazer Press, 1985).
Using a biodynamic solar greenhouse for saving money on crops and livestock.

Poisson, Leandre, and Gretchen Vogel Poisson, *Solar Gardening* (White River Junction, VT: Chelsea Green Publishing, 1994).
Grow vegetables year-round with "mini-greenhouses."

Smith, Shane, *The Bountiful Solar Greenhouse* (Santa Fe, NM: John Muir Publications, 1982).
Growing food year-round with solar greenhouses.

Herbs and Flowers
Byczynski, Lynn, *The Flower Farmer* (White River Junction, VT: Chelsea Green Publishing, 1997).
Guide to raising and selling organic cut flowers.

Miller, Richard Alan, *The Potential of Herbs as a Cash Crop,* 2nd ed. (Metairie, LA: Acres USA, 1985).
The classic guide to the herb business.

Shores, Sandy, *Growing and Selling Fresh-Cut Herbs* (Pownal, VT: Storey Books, 1999).
Excellent handbook for herbal agripreneurs.

Stevens, Alan, *Field-Grown Cut Flowers* (Edgerton, WI: Avatar's World, 1997).
Production guide for fresh and dried cut flowers.

Sturdivant, Lee, *Herbs for Sale* (Friday Harbor, WA: San Juan Naturals, 1994).
Growing and marketing herbs.

———, and Tim Blakley, *Medicinal Herbs in the Garden, Field, and Marketplace* (Friday Harbor, WA: San Juan Naturals, 1998).
Guide to establishing a medicinal herb business.

History
Cochrane, Willard W., *The Development of American Agriculture: A Historical Analysis* (Minneapolis: University of Minnesota Press, 1981).
A look at how we got to where we are in agriculture.

Homesteading

Emery, Carla, *The Encyclopedia of Country Living*, 9th ed. (Seattle, WA: Sasquatch Books, 1994).
The best general information book I have found, from winemaking to meat drying.

Wigginton, Eliot, *The Foxfire Book*, vol. 1 (Garden City, NY: Anchor Press, 1972).
Homecraft how-to. Nine other volumes followed this one.

Honeybees

Bonney, Richard E., *Beekeeping: A Practical Guide* (Pownal, VT: Storey Books, 1993).
Acquiring and managing bees.

Morse, Roger A., *The New Complete Guide to Beekeeping* (Woodstock, VT: Countryman Press, 1994).
Starting and maintaining bees.

Inspirational

Berry, Wendell, *The Gift of Good Land* (Sisters, OR: Farmer's Book Service, 1981).
More cultural and agricultural essays.

———, *The Unsettling of America* (New York: Avon Books, 1978).
Essays on the importance of sustainable agriculture.

Logsdon, Gene, *You Can Go Home Again* (White River Junction, VT: Chelsea Green, 1998).
More wonderful inspiration.

Machinery and Computers

Bowman, Greg, ed., *Steel in the Field* (Burlington, VT: Sustainable Agriculture Network, 1997).
An excellent guide to weed management tools.

Campidonica, Mark, *How to Find Agricultural Information on the Internet* (Oakland: University of California DANR, 1997).
Good introduction to finding farming resources on the Net.

Farm Conveniences and How to Use Them (New York: Lyons Press, 1999).
Labor-saving devices from the 1800s.

Marketing Methods

Ableman, Michael, *On Good Land* (San Francisco: Chronicle Books, 1998).
Story of a small community farm in California.

Copeland, John D., *Recreational Access to Private Lands*, 2nd ed. (Fayetteville, AR: National Center for Agricultural Law Research and Information, 1998).
Legal advice on establishing on-farm activities.

Doane, D. Howard, *Vertical Farm Diversification* (Norman: University of Oklahoma Press, 1950).
Taking farms beyond raw material production.

Groh, Trauger, and Steven McFadden, *Farms of Tomorrow Revisited* (White River Junction, VT: Chelsea Green Publishing, 1997).
A blueprint for Community Supported Agriculture farms.

Rogak, Lisa, *The Complete Country Business Guide* (Grafton, NH: William Hills Publishing, 1998).
Introduction to starting a rural business.

Pastures and Cover Crops
Hughes, H.D., Maurice E. Heath, and Darrel S. Metcalfe, *Forages,* 2nd ed. (Ames: Iowa State University Press, 1962).
Excellent guide to forages. The 5th edition (1995) is a two-volume set.

Managing Cover Crops Profitably, 2nd ed. (Burlington, VT: Sustainable Agriculture Network, 1998).
An excellent guide to choosing and growing cover crops.

Murphy, Bill, *Greener Pastures on Your Side of the Fence* (Colchester, VT: Arriba Publishing, 1987).
Using the Voisin system of grazing management to improve pasture productivity.

Nation, Allan, *Quality Pasture* (Jackson, MS: Green Park Press, 1995).
Creating, managing, and profiting from quality pasture.

Smith, Burt, Pingsun Leung, and George Love, *Intensive Grazing Management* (Kamuela, HI: The Graziers Hui, 1986).
Managing forage and animals for profit.

Poultry
The American Standard of Perfection (Troy, NY: American Poultry Association, Inc., 1993).
A complete description of all recognized breeds and varieties of domestic poultry, periodically revised.

Damerow, Gail, *A Guide to Raising Chickens* (Pownal, VT: Storey Publishing, 1995).
An in-depth guide.

Hastings Belshaw, R.H., *Guinea Fowl of the World* (Northamptonshire, UK: Nimrod Book Services, 1985).
Guide to breeding guineas.

Lee, Andy, and Pat Foreman, *Chicken Tractor, Straw Bale Edition* (Columbus, NC: Good Earth Publications, 1998).
Portable chicken pens in the garden.

Levi, Wendell Mitchell, *The Pigeon* (Sumter, SC: Levi Publishing Co., Inc., 1957).
Complete information on raising pigeons.

Salatin, Joel, *Pastured Poultry Profits* (Swoope, VA: Polyface Farms, 1993).
Rotational grazing chickens in portable cages.

Schwanz, Lee, ed., *The Family Poultry Flock* (Brookfield, WI: Farmer's Digest, Inc., 1981).
Basics of choosing and raising poultry.

Thear, Katie, *Free-range Poultry* (Ipswich, UK: Farming Press Books, 1990).
Guide to grazing chickens free-range.

Shelter and Fencing
Burch, Monte, *How to Build Small Barns and Outbuildings* (Pownal, VT: Storey Books, 1992).
Fundamentals of general construction.

Damerow, Gail, *Fences for Pasture and Garden* (Pownal, VT: Storey Books, 1992).
Selecting, planning, and building fences.

Merrilees, Dong, Ralph Wolfe, and E. Loveday, *Low-Cost Pole Building Construction* (Pownal, VT: Storey Books, 1980).
How to build a small home, barn or other pole structure.

Small-Acreage Farming
Angier, Bradford, *One Acre and Security* (New York: Random House, 1972).
General information on starting a farm.

Bromfield, Louis, *From My Experience* (New York: Harper & Brothers, 1955).
More about Malabar.

———, *Malabar Farm* (New York: Ballantine, 1947, 1970).
The author returns to organic farming and tells about it.

Logsdon, Gene, *The Contrary Farmer* (White River Junction, VT: Chelsea Green, 1993).
Great inspiration and useful advice.

Miller, Ralph C. and Lynn R. Miller, eds. *Ten Acres Enough: The Small Farm Dream Is Possible* (Sisters, OR: Farmer's Book Service, 1981).
A reprint of the 1864 classic, with updated essays. Part inspiration, part useful advice.

Olson, Michael, *MetroFarm* (Santa Cruz, CA: TS Books, 1994).
Another good guide to small parcel success.

Salatin, Joel, *You Can Farm* (Swoope, VA: Polyface Farms, 1998).
The author explains his system for success.

Smart, Charles Allen, *RFD* (Athens: Ohio University Press, 1998).
City dweller turns farmer during the Depression.

Smith, Miranda, ed., *The Real Dirt* (Burlington, VT: Northeast Region SARE, 1994).
Organic and low-input practices in the Northeast.

Whateley, Booker T., *How to Make $100,000 Farming 25 Acres* (Chillicothe, IL: American Botanist, 1996).
The classic on small farm success.

The Yearbooks of Agriculture, 1910–1962, 1976. (Washington, DC: US Department of Agriculture).
These books have all sorts of useful information. In particular, I recommend 1938: Soils and Men; 1941: Climate and Man; 1948: Grass; 1949: Trees; 1955: Water; 1957: Soil; and 1976: Living on a Few Acres.

Small Fruits and Tree Crops

Harlan, Michael, and Linda Harlan, *Growing Profits* (Citrus Heights, CA: Moneta Publications, 1997).
Starting and operating a backyard nursery.

Otto, Stella, *The Backyard Berry Book* (Maple City, MI: OttoGraphics, 1995).
How to grow berries.

———, *The Backyard Orchardist* (Maple City, MI: OttoGraphics, 1993).
Growing fruit trees in the home garden.

Wampler, Ralph L., and James E. Motes, *Pick-Your-Own Farming* (Norman: University of Oklahoma Press, 1984).
Growing crops for U-pick farms.

Small Stock

Cheeke, Peter R., Nephi M. Patton, and George S. Templeton, *Rabbit Production*, 5th ed. (Danville, IL: Interstate Printers and Publishers, Inc., 1982).
Excellent guide to raising rabbits.

Drummond, Susan Black, *Angora Goats the Northern Way* (Freeport, MI: Stoney Lonesone Farm, 1985).
Raising and utilizing angora goats.

Klober, Kelly, *A Guide to Raising Pigs* (Pownal, VT: Storey Books, 1997).
An excellent source for small-scale pig raising.

Kruesi, William K., *The Sheep Raiser's Manual* (Charlotte, VT: Williamson Publishing, 1985).
Good guide to raising sheep.

MacKenzie, David, *Goat Husbandry*, 4th ed. (London: Faber & Faber Limited, 1980).
Guide to raising goats.

Thornton, Keith, *Outdoor Pig Production* (Ipswich, UK: Farming Press Books, 1990).
Basics of raising pigs.

Soils

Albrecht, William A., *The Albrecht Papers*, Vol. 1, (Metairie, LA: Acres USA, 1975).
Classic research on sustainable soils and their importance. Three more volumes follow.

Buckman, Harry O., and Nyle C. Brady, *The Nature and Properties of Soils*, 7th ed.
(New York: Macmillan Company, 1969).
An excellent guide to soils and soil management.

Faulkner, Edward H., *Plowman's Folly* (New York: Grosset & Dunlap, 1943).
Alternative cultivation methods.

Gershuny, Grace, *Start with the Soil* (Emmaus, PA: Rodale, 1993).
Organic gardener's guide to improving soil.

————, and Joseph Smillie. *The Soul of Soil*, 3rd ed. (White River Junction, VT: Chelsea
Green, 1996).
A guide to ecological soil management.

Kinsey, Neal, and Charles Walters, *Hands-On Agronomy* (Metairie, LA: Acres USA,
1993).
Building and maintaining the soil.

Walters, Charles Jr., and C.J. Fenzau, *An Acres USA Primer* (Metairie, LA: Acres USA,
1979).
Principles of soil care from an unusual perspective.

Sustainable Practices

Altieri, Miguel A., *Agroecology* (Boulder, CO: Westview Press, 1987).
A scientific justification of alternative agriculture.

Avory, Allan, *Holistic Resource Management* (Covelo, CA: Island Press, 1988).
Comprehensive system planning.

Edwards, Clive, et al., eds. *Sustainable Agricultural Systems* (Ankeny, IA: Soil and Water
Conservation Society, 1990).
Practical research on sustainable agriculture.

Fukuoka, Masanobu, *The Natural Way of Farming* (New York: Japan Publications,
Inc., 1985).
"Do-nothing" organic farming in Japan.

Jackson, Wes, et al., eds., *Meeting the Expectations of the Land* (Berkeley, CA: North
Point Press, 1984).
Essays on sustainable agriculture.

————, *New Roots for Agriculture* (Lincoln: University of Nebraska Press, 1985).
Reasons for sustainable agriculture and how to practice it.

National Research Council, *Alternative Agriculture* (Washington, DC: National Academy Press, 1989)
Alternative farming benefits and case studies.

Vegetables

Damerow, Gail, *The Perfect Pumpkin* (Pownal, VT: Storey Books, 1997).
Growing and using pumpkins.

Deppe, Carol, *Breed Your Own Vegetable Varieties* (Boston: Little, Brown and Company, 1993).
Plant breeding for the home gardener.

DeWitt, Dave, and Paul W. Bosland, *The Pepper Garden* (Berkeley, CA: Ten Speed Press, 1993).
How to grow peppers.

Jeavons, John, *How to Grow More Vegetables* 5th ed. (Berkeley, CA: Ten Speed Press, 1995).
Sustainable biointensive organic horticulture.

Lorenz, Oscar A., and Donald N. Maynard, *Knott's Handbook for Vegetable Growers*, 3rd ed. (New York: John Wiley & Sons, 1988).
The essential authority on production of vegetables.

Stamets, Paul, *Growing Gourmet and Medicinal Mushrooms* (Berkeley, CA: Ten Speed Press, 1993).
A bible of mushrooms.

Weaver, William Woys, *Heirloom Vegetable Gardening* (New York: Henry Holt and Company, 1997).
Planting, growing, and saving seeds.

Wilbur, Charles W., *How to Grow World Record Tomatoes* (Metairie, LA: Acres USA, 1998).
Growing huge organic tomato plants.

Book Sources

Acres USA, PO Box 8800, Metairie, LA 70011; 1-800-355-5313.
Publishes and carries a variety of farming books.

agAccess, 603 4th Street, Davis, CA 95616; 530-756-7177.
Publishes several farming books.

Chelsea Green Publishing Company, PO Box 428, White River Junction, VT 05001; 1-800-639-4099.
Publishes and carries a variety of farming books.

Country Store Books, 307 E. Ash, PMB 57, Columbia, MO 65201; fax: 573-442-9887.
Carries a variety of farming books.

Countryman Press, PO Box 175, Woodstock, VT 05091-0175; 1-800-245-4151.
Focuses on the outdoors and gardening.

Farmers Book Service, *Small Farmers Journal,* PO Box 1627, Sisters, OR 97759; 541-549-2064.
Publishes and carries a variety of farming books.

Hancock House Publishers, 1431 Harrison Ave., Blaine, WA 98230-5005; 1-800-938-1114, http://www.hancockwildlife.org.
Publishes aviculture (bird) books.

Lessiter Publications Inc., PO Box 624, Brookfield, WI 53008-0624.
Publishes books focusing on horses and smithing.

San Juan Naturals, PO Box 642, Friday Harbor, WA 98250; 1-800-770-9070.
Publishes herbal books.

Storey Publishing Inc., PO Box 445, Pownal, VT 05261; 1-800-441-5700, http://www.storey.com.
Publishes a wide variety of farming books.

Sustainable Agriculture Network, Hills Building, University of Vermont, Burlington, VT 05405-0082; http://www.sare.org.
Publishes several sustainable farming books.

Ten Speed Press, PO Box 7123, Berkeley, CA 94707; 510-559-1600.
Publishes a variety of garden books.

Sources for Out-of-Print Books
Country Books, 826 Power Line Road, Paris, Ontario N3L 3E3, Canada.

Little Creek Bookshop, PO Box 100, Vass, NC 28394.

Marvin Jager, Washington, NH 03280.

Robert Gear, Box 1137, Greenfield MA 01302; 413-337-4844.

WXICOF, 914 Riske Lane, Wentzville, MO 63385; 636-828-5100.
Large collection of foreign farming books.

Periodicals

Magazines and Newspapers
Magazines about specific crops or animals offer the best opportunity to further your knowledge. They not only have up-to-date information on what you wish to raise,

but they also contain display advertisements and breeder's directories that list places to buy what you need and information contacts.

Acres USA, PO Box 8800, Metairie, LA 70011-8800; 504-889-2100. Monthly.
Eco-agriculture information, ranging from standard to extreme.

Alpacas, PO Box 1992, Estes Park, CO 80517-19923; 970-586-5357. Quarterly.
Information on alpacas.

American Bee Journal, 51 S. 2nd Street, Hamilton, IL 62341; 217-847-3324. Monthly.
Information on honeybees.

Animal Finders' Guide, PO Box 99, Prairie Creek, IN 47869;
812-898-2678; http://www.animalfindersguide.com. 18 issues per year.
Listings of rare and exotic animals for sale.

Back Home, PO Box 70, Hendersonville, NC 28793; 828-696-3838. Bimonthly.
Sustainable living.

Bee Culture, The A.I. Root Co., 623 W. Liberty Street, Medina, OH 44256; 1-800-289-7668. Monthly.
American beekeeping.

Biodynamics, PO Box 29135, San Francisco, CA 94129-0135;
415-561-7797; http://www.biodynamics.com. Bimonthly.
Information on the biodynamic (organic) method of agriculture.

The Brayer, American Donkey & Mule Society Inc., 2901 N. Elm Street, Denton, TX 76201; 940-382-6845. Bimonthly.
Donkey and mule information.

Country Garden & Smallholding, Buriton House, Station Road, Newport, Saffron Walden, Essex CB11 3PL, United Kingdom. Monthly.
Small farming in England.

Countryside & Small Stock Journal, W11564 Hwy. 64, Withee, WI 54498; 1-800-551-5691. Bimonthly.
Homesteading.

Domestic Rabbits, Box 426, Bloomington, IL 61702; 309-664-7500. Bimonthly.
News from the American Rabbit Breeders Association.

Draft Horse Journal, PO Box 670, Waverly, IA 50677; 319-352-4046. Quarterly.
Draft horse news.

Fish Farming News, PO Box 37, Stonington, ME 04681; 207-367-2396. 7 issues per year.
Information on the aquaculture business.

Goat Rancher, 731 Sandy Branch Road, Sarah, MS 38665; 601-562-9529. Monthly.
Information on meat goats.

The Homestead Connection, 721 McKinnon Road, Boston, GA 31626; 912-228-4215. Bimonthly.
Focuses on goats.

Llamas, PO Box 250, Jackson, CA 95642; 209-295-7800. 5 issues per year.
Information on camelids.

The Maine Organic Farmer & Gardener, PO Box 2176, Augusta, ME 04338; 207-622-3118. Quarterly.
Organic agriculture.

Mother Earth News.
I am actually recommending the old issues, pre-1986, for useful information on small farm homesteading. The later issues are more environmentally focused, and have less information applicable to farmers.

Mules and More, PO Box 460, Bland, MO 65014; 573-646-3934. Monthly.
Information on mules.

National Poultry News, PO Box 1647, Easley, SC 296411647; 864-855-0140. 5 issues per year.
Promotes poultry.

The Natural Farmer, c/o NOFA, 411 Sheldon Road, Barre, MA 01005; 508-355-2853. Quarterly.
Organic farming information.

North American Elk, 1708 N. Prairie View Road, Platte City, MO 64079; 816-431-3605. Bimonthly.
Elk information.

Organic Gardening, 33 E. Minor Street, Emmaus, PA 18098; 610-967-5171. Monthly.
Growing chemical-free food and flowers.

Poultry Press, PO Box 542, Connersville, IN 47331-0542; 765-827-0932. Monthly.
Promotes standardbred poultry.

Ranch & Rural Living, PO Box 2678, San Angelo, TX 76902; 915-655-4434. Monthly.
Focuses on sheep and goats.

Rare Breeds Journal, PO Box 66, Crawford, NE 69339; 308-665-1431. Bimonthly.
Exotic and minor breeds of animals for the pet industry.

Rural Heritage, 281 Dean Ridge Lane, Gainesboro, TN 38562-5039; 931-268-0655; http://www.ruralheritage.com. Bimonthly.
Focuses on farming and logging with draft animals.

The Ranch Dog Trainer, HC 69 Box 300, Oscar, OK 73569; 580-437-2215. Bimonthly.
Information on working stockdogs.

Rural Property Bulletin, PO Box 608, Valentine, NE 69201; 402-376-2617. Monthly.
Rural property sale listings.

Small Farmer's Journal, PO Box 1627, Sisters, OR 97759; 541-549-2064. Quarterly.
Small farming, focusing on draft animals.

Small Farm Today, 3903 W. Ridge Trail Rd, Clark, MO 65243; 1-800-633-2535.
http://www.smallfarmtoday.com. Bimonthly.
How-to information on alternative and traditional crops, livestock, and direct marketing.

The Stockman Grass Farmer, PO Box 2300, Ridgeland, MS 39158-2300;
1-800-748-9808. Monthly.
Information on intensive rotational grazing.

Town and Country Farmer, PO Box 798, Benalla, Victoria, Australia, 3672. Quarterly.
Family farm magazine from Australia.

Western Mule, PO Box 46, Marshfield, MO 65706; 417-859-6853. Monthly.
Information on mules.

Wildlife Harvest, PO Box 96, Goose Lake, IA 52750; 319-259-4000. Monthly.
Information on gamebird production and hunting.

Wings & Hooves, Route 1, Box 32, Forestburg, TX 76239-9706;
940-964-2314. Monthly.
Exotic animals.

Newsletters
Newsletters are another source of information, ranging from the magazine-like
Growing for Market to the university-published *Sustainable Agriculture.*

Ag News & Views, The Samuel Roberts Noble Foundation, PO Box 2180, Ardmore,
OK 73502-2180; 580-223-5810, http://www.noble.org. 8 pages. Monthly.
General farm information.

Alternative Agriculture News, 9200 Edmonston Road, Suite 117, Greenbelt, MD
20770-1551. 4 pages. Published by Henry A. Wallace Institute for Alternative
Agriculture. One-year subscription with membership. Monthly.
Sustainable news and calendar.

American Livestock Breeds Conservancy News, PO Box 477, Pittsboro, NC 27312;
919-542-5704. 12 pages. One-year subscription with membership. Bimonthly.
Information on rare breeds.

Beginning Farmer News and Notes, PO Box 736, Hartington, NE 68739. 2 pages.
Published by the Center for Rural Affairs. One-year subscription. Monthly.
News and inspiration.

BioOptions, 352 Alderman Hall, 1970 Folwell Avenue, St. Paul, MN 55108. 16 pages. Published by the Center for Alternative Plant and Animal Products. One-year subscription. Quarterly.
News and information on alternatives.

Center for Rural Affairs, PO Box 406, Walthill, NE 68067; 402-846-5428; http://www. cfra.org. 6 pages. One-year subscription. Monthly.
Information on national events affecting family farms.

The Cut Flower Quarterly. MPO Box 268, Oberlin, OH 44074; http://www.ascfg.org. 24 pages. Published by the Association of Specialty Cut Flower Growers. One-year subscription. Quarterly.
Field and greenhouse cut-flower information.

Growing for Market, PO Box 3747, Lawrence, KS 66046; 785-748-0605. 20 pages. One-year subscription. Monthly.
News and ideas for market gardeners.

HortIdeas, 460 Black Lick Road, Gravel Switch, KY 40328. 12 pages. One-year subscription. Monthly.
Reports the latest research and tools for gardeners.

Leopold Letter, 209 Curtiss Hall, Iowa State University, Ames, IA 50011-1050; 515-294-3711; http://www.leopold.iastate.edu/. 12 pages. Published by the Leopold Center for Sustainable Agriculture. One-year subscription. Quarterly.
News and research on sustainable agriculture.

Small Farm News, Small Farm Center, One Shields Avenue, University of California, Davis, CA 95616-8699; 530-7528136. 10 pages. Bimonthly.
Specialty crops and small farm news.

Sustainable Agriculture, 405 Coffey Hall, 1420 Eckles Avenue, University of Minnesota, St. Paul, MN 55108; 612-625-1794. 4 pages. Bimonthly.
Sustainable agriculture news.

Internet Sites

There are several Internet addresses scattered throughout the lists above. I have provided some more starting points here. There are many more sites, of course, so do some keyword searches (on several search engines) using topics of interest (e.g., sustainable, buffalo, organic, small farm). Use links from the sites you find to explore more options. If you are new to the Net and need a good starting point, go to the Animal Science/Oklahoma State site.

AgEBB/University of Missouri: http://www.ext.missouri.edu/agebb/.
Farming information.

Animal Science/Oklahoma State: http://www.ansi.okstate.edu/.

CityFarmer: http://www.cityfarmer.org/.
Urban farming information.

Cyberfarm: http://www.ag.ohio-state.edu/~farm/.
Small farm information.

Forage Information System: http://forages.orst.edu/.
Forage information.

Greenweb: http://www.boldweb.com/greenweb.htm.
Gardening information.

IPM Almanac: http://www.ipmalmanac.com.
Information on Integrated Pest Management.

Langston University: http://www.luresext.edu/.
Aquaculture and goat information.

MachineFinder: http://www.machinefinder.com.
Search for used equipment on-line.

MISA: http://www.misa.umn.edu/.
Sustainable agriculture information.

North Dakota Extension: http://www.ext.nodak.edu/extpubs/.
Farming information.

Open Air Farmers: http://www.openair.org/.
Farmers' market information.

Ostriches Online: http://www.ostrichesonline.com/general.
Includes on-line newsletter.

Rural Estate Network: http://www.ruralspace.com/.
Property for sale.

In addition to the Internet, there are also Newsgroups and Mailing Lists. Newsgroup availability varies from server to server, so I recommend just looking at your available list for farm-related groups. Mailing Lists are basically digests of messages sent to you via e-mail. Two Mailing Lists I like are:

Graze-L, providing information on rotational grazing and forages. To subscribe, send an e-mail to Majordomo@taranaki.ac.nz, with the sentence "subscribe graze-l-digest" in the message body.

Sheep-L, providing information on sheep. To subscribe, send an e-mail to listserv@listserv.uu.se, with the sentence "SUBSCRIBE SHEEP-L" in the message body.

University Sources

Consult your local extension office and university agriculture department and ask about any sustainable, value-added, or small farm programs it has available. Check if there are any extension specialists in topics you are investigating. My list here is of universities and departments that I feel are promoting small farm issues or have useful small farm information.

California

Small Farm Center, One Shields Avenue, University of California, Davis, CA 95616-8699; 530-752-8136; http://www.sfc.ucdavis.edu.

University of California, Division of Agriculture and Natural Resources, Communication Services, 6701 San Pablo Avenue, Oakland, CA 94608-1239; 1-800-994-8849.

Colorado

Cooperative Extension, Colorado State University, Fort Collins, CO 80523; 970-491-6198.

Florida

Farm Publications, Perry-Paige Bldg. Room 202C, Florida A&M University, Tallahassee, FL 32307; 904-599-3547.

Idaho

Corrine Lyle, College of Agriculture, University of Idaho, Moscow, ID 83843; 208-885-5883. *New-farmer packets.*

Indiana

Agricultural Communications Service (Purdue University), Media Distribution Center, 301 S. 2nd Street, Lafayette, IN 47901-1232; 1-800-398-4636.

Iowa

Leopold Center for Sustainable Agriculture, 209 Curtiss Hall, Iowa State University, Ames, IA 50011-1050; 515-294-3711, http://www.leopold.iastate.edu/.

Louisiana

Publications Office, Ag Center Communications (Louisiana State University), PO Box 25100, Baton Rouge, LA 70894-5100.

Maryland

Frederick County Cooperative Extension Service, University of Maryland, 330 Montevue Lane, Frederick, MD 21702; 301-694-1594; http://www.agnr.umd.edu/users/frederick/pubs. *Fact sheets.*

Michigan
Michigan State University Bulletin Office, 10-B Agriculture Hall, East Lansing, MI 48824-1039; 517-355-0240.

Minnesota
Center for Alternative Plant and Animal Products, 352 Alderman Hall, 1970 Folwell Avenue, St. Paul, MN 55108.

Missouri
Missouri Alternatives Center, 531 Clark Hall, University of Missouri, Columbia, MO 65211; 314-882-1905. Packets — free to Missouri residents.

Montana
Extension Publications, Montana State University, PO Box 172040, Bozeman, MT 59717-2040; 406-994-3273.

New York
Farming Alternatives Program, Warren Hall, Cornell University, Ithaca, NY 14853; 607-255-9832.

Pennsylvania
Publications Distribution Center, The Pennsylvania State University, 112 Ag. Admin. Bldg., University Park, PA 16802-2602; 814-865-6713, http://Agalternatives.cas.psu.edu.
Fact sheets with budgets.

West Virginia
West Virginia University Extension Service, Communications and Educational Tech, 810 Knapp Hall, PO Box 6031, Morgantown, WV 26506-6031.

Wisconsin
Cooperative Extension Publications (University of Wisconsin), Room 170, Dept. GV, 630 W. Mifflin Street, Madison, WI 53703; 608-262-3346.

Federal Agencies

In addition to the agencies below, I recommend contacting your state Department of Agriculture to find out what offices and grants could aid your farm. For example, in Missouri we have the AgriMissouri program for adding value and labeling. Your State Department of Agriculture is also your best source for legal information.

Alternative Farming Systems Information Center, National Agricultural Library Room 304, 10301 Baltimore Avenue, Beltsville, MD 20705-2351; 301-504-6559, http://www.nal.usda.gov/afsic/.
Publishes information on farming alternatives.

Appropriate Technology Transfer for Rural Areas (ATTRA), PO Box 3657, Fayetteville, AR 72702; 800-346-9140; http://www. attra.org.
Provides free information on sustainable agriculture topics.

Small Business Center, U.S. Chamber of Commerce, 1615 H Street NW, Washington, DC 20062; 202-463-5503.
Publishes the Small Business Financial Resource Guide. *Has addresses of Small Business Development Centers and other programs.*

Small Farm Program, USDA-CSREES, Stop 2220, 1400 Independence Avenue SW, Washington, DC 20250-2220; 1-800-583-3071.
Publications, factsheets.

Sustainable Agriculture Research and Education (SARE), 1400 Independence Avenue SW, Room 3851 South Bldg., Washington, DC 20250-1910; 202-720-5203, http://www.sare.org.
Research and grants on sustainable agriculture. The U.S. has four regional offices.

Map Sources

There are several sources for maps of your property. Cartographic maps and platte books may be purchased from your local USDA Natural Resources Conservation Service (NRCS) agency or may be found commercially. Aerial photographic maps may be purchased from your local Farm Service Agency. Soil maps may be found at a local or state soil survey office. If you cannot find where these are located, here are national offices:

Farm Service Agency, Aerial Photography Field Office, 2222 W. 2300 S, Salt Lake City, UT 84119-2020; 801-975-3500; http://www.fsa.usda.gov/.

National Soil Survey Center, Federal Bldg. Room 152, 100 Centennial Mall N, Lincoln, NE 68508-3866; 402-437-5499; http://www.statlab.iastate.edu/soils/nssc/.

National Cartography and Geospatial Center Library, PO Box 6567, Fort Worth, TX 76115-0567; 817-509-3394; http://www.ftw.nrcs.usda.gov/ncg/ncg.html.

Resource Lists

Barton, Barbara J., *Gardening by Mail* (Sebastapol, CA: Tusker Press, 1994).
More than 1,000 nurseries and seed sources.

Directory of Flower & Herb Buyers. Prairie Oak Seeds, PO Box 382, Maryville, MO 64468; 660-562-3743.
A small list of buyers, organized by state.

Facciola, Stephen, *Cornucopia II* (Vista, CA: Kampong Publications, 1999).
Thousands of edible plants and their sources.

Gardener's Source Guide, 1998. GSG, PO Box 206, Gowanda, NY 14070-0206.
More than 900 mail-order sources for gardening.

A Guide to USDA and Other Federal Resources for Sustainable Agriculture and Forestry Enterprises, 1998. U.S. Department of Agriculture with the Michael Fields Agricultural Institute.
List of federal programs supporting sustainable agriculture.

The Herbal Green Pages. Herb Growing and Marketing Network, PO Box 245, Silver Spring, PA 17575-0245; 717-393-3295; http://www.herbnet.com.
More than 6,000 herb-related businesses, associations, and suppliers listed.

McRae, Bobbi A., *The New Fiberworks Sourcebook* (Austin, TX: Fiberworks Publications, 1993).
Thousands of sources for fibers and fiber manufacturing equipment.

The National Organic Directory. Community Alliance with Family Farmers, PO Box 363, Davis, CA 95617-9900; 800-852-3832; http://www.caff.org.
Over 1,000 organic farmers, wholesalers, suppliers, and certification groups listed.

The Organic Pages. Organic Trade Association, PO Box 1078, Greenfield, MA 01302; 413-774-7511; http://www.ota.com.
More than 1,000 organic growers, associations, suppliers, and certification groups listed.

Skolnick, Solomon M., *The Home Gardener's Source* (New York: Random House, 1997).
Hundreds of listings by categories.

Small Farm Resource Guide, 1998. Small Farm Program, USDA-CSREES Stop 2220, 1400 Independence Ave SW, Washington, DC 20250-2220; 202-401-4385; http://www.reeusda.gov/smallfarm.
Useful lists of small farm programs and resources by state.

Sustainable Agriculture Directory of Expertise, 3rd ed. (Burlington, VT: Sustainable Agriculture Network, 1996).
Lists more than 700 organizations and individuals with sustainable knowledge.

Seed and Plant Catalogs

There are hundreds of seed companies. These are some that I like:

Abundant Life Seed Foundation, PO Box 772, Port Townsend, WA 98368.
Open-pollinated, rare, and heirloom seeds.

Albert Lea Seed House Inc., PO Box 127, Albert Lea, MN 56007; 1-800-353-5247.
Field crop and pasture seed.

Ames' Orchard & Nursery, 18292 Wildlife Road, Fayetteville, AR 72701; 501-443-0282.
Apples and other orchard trees.

Bountiful Gardens, 18001 Shafer Ranch Road, Willits, CA 95490; 707-459-6410; e-mail: bountiful@zapcom.net.
Organic vegetables, cover crops, and information.

E & R Seed, 1356 E. 200 S, Monroe, IN 46772.
Vegetables, flowers, and supplies.

Fedco Seeds, PO Box 520, Waterville, ME 04903.
Vegetables, herbs, flowers, cover crops. Tree and bulb catalog also available.

Filaree Farm, 182 Conconully Highway, Okanogan, WA 98840-9774; 509-422-6940.
Garlic and supplies.

Fungi Perfecti, PO Box 7634, Olympia, WA 98507; 1-800-780-9126.
Gourmet and medicinal mushrooms and supplies.

Garden City Seeds, 778 Highway 93 N., Hamilton, MT 59840; 406-961-4837; e-mail: seeds@juno.com.
Vegetables and supplies.

Harris Seeds, 60 Saginaw Drive, PO Box 22960, Rochester, NY 14692-2960; 716-442-0410.
Vegetables and flowers.

Johnny's Selected Seeds, Foss Hill Road, Albion, ME 04910-9731; 207-437-4395.
Vegetables for market growers.

Jordan Seeds Inc., 6400 Upper Afton Road, Woodbury, MN 55125; 612-738-3422.
Vegetables.

J.W. Jung Seed Co., 335 S. High Street, Randolph, WI 53957-0001; 1-800-297-3123.
Vegetables, flowers, and berries.

Oregon Exotics Nursery, 1065 Messinger Road, Grants Pass, OR 97527; 541-846-7578.
Foreign and exotic fruits and vegetables.

Otis S. Twilley Seeds, PO Box 65, Trevose, PA 19047; 215-639-8800.
Vegetables and flowers.

Peaceful Valley Farm Supply, PO Box 2209, Grass Valley, CA 95945; 916-272-4769.
Organic farming tools and supplies.

Richters, Goodwood, Ontario L0C 1A0, Canada; 905-640-6677; http://www.richters.com.
Herbs and books.

Ronniger's Seed & Potato Co., PO Box 307, Ellensburg, WA 98926; 1-800-846-6178.
Potatoes, garlic, and onions.

Rupp Seed, 17919 County Road B, Wauseon, OH 43567; 419-337-1841.
Vegetables, pumpkins, and watermelons.

Sandhill Preservation Center, 1878 230th Street, Calamus, IA 52729.
All heirloom, untreated. Farmer owned. Heirloom vegetables and poultry.

Seed Savers Exchange, 3076 N. Winn Road, Decorah, IA 52101.
Farmer seed exchange catalogs and yearbook.

Seeds of Change, PO Box 15700, Santa Fe, NM 87506-5700; 1-888-762-7333;
http://www.seedsofchange.com.
Organic vegetables, herbs, and supplies.

Shepherd's Garden Seeds, 30 Irene Street, Torrington, CT 06790; 860-482-3638.
Flowers, herbs, and vegetables.

R.H. Shumway's, PO Box 1, Graniteville, SC 29829-0001; 803-663-9771.
Vegetables and more. HPS (commercial) and regular catalogs available.

Southern Exposure Seed Exchange, PO Box 170, Earlysville, VA 22936; 804-973-4703.
Farmer owned. Vegetables.

Stark Bros., Box 10, Louisiana, MO 63353-0010; 1-800-325-4180.
Fruit trees and landscaping.

Territorial Seed Company, PO Box 157, Cottage Grove, OR 97424; 541-942-9547.
Vegetables plus.

Tomato Growers Supply Company, PO Box 2237, Fort Myers, FL 33902; 941-768-1119.
Tomatoes and peppers only.

Totally Tomatoes, PO Box 1626, Augusta, GA 30903-1626;
803-663-0016.
Tomatoes and peppers only.

Supplies

A.H. Hummert Seed Co., 2746 Chouteau Avenue, St. Louis, MO 63103; 314-506-4500.
Horticulture supplies.

A.M. Leonard Inc., 241 Fox Drive, Piqua, OH 45356; 1-800-543-8955;
http://www.amleo.com.
Garden supplies.

Agronics, 800 Madison NE, Albuquerque, NM 87110.
Soil mineralization supplies.

CropKing Inc., 5050 Greenwich Road, Seville, OH 44273; 330-769-2002.
Hydroponic greenhouse supplies.

Cumberland General Store, 1 Highway 68, Crossville, TN 38555; 800-334-4640.
Homesteading equipment.

Denman & Company, 1202 E. Pine Street, Placentia, CA 92670; 714-524-0668.
Wheel hoes and garden hand tools.

Gardens Alive!, 5100 Schenley Place, Lawrenceburg, IN 47025; 812-537-8651.
Organic fertilizers and supplies.

Gardener's Supply Company, 128 Intervale Road, Burlington, VT 05401; 1-800-863-1700; http://www.gardeners.com.
Garden supplies.

Gempler's, PO Box 270, Belleville, WI 53508; 1-800382-8473;
http://www.gemplers.com.
Supplies of all sorts, including IPM.

Harmony Farm Supply, PO Box 460, Graton, CA 95444;
707-823-9125.
Organic crop and garden supplies.

Lehman's, PO Box 41, Kidron, OH 44636; 330-857-1111; http://www.lehmans.com.
Homesteading equipment.

Midwest Plan Service, 1222 Davidson Hall, Iowa State University, Ames, IA 50011-3080; 1-800-562-3618.
Livestock shelter plans.

Nasco, 901 Janesville Avenue, Fort Atkinson, WI 53538-0901; 1-800-558-9595.
Farm and ranch supplies.

Natural Solutions, 2218 Sunswept Court, Sugar Land, TX 77478; 800-231-9725.
Organic supplies.

Ohio Earth Foods, 5488 Swamp Street NE, Hartville, OH 44632; 330-877-9356.
Natural fertilizer and supplies.

Peaceful Valley Farm Supply, PO Box 2209, Grass Valley, CA 95945; 916-272-4769.
Natural supplies, including irrigation.

Planet Natural, PO Box 3146, Bozeman, MT 59772; 406-587-0223;
http://www.planetnatural.com.
Organic and natural garden supplies.

Produce Promotions, 19 S. Water Market, Chicago, IL 60608; 312-633-9821.
Road signs, price cards, displays for produce.

Prohoe, PO Box 87, Munden, KS 66959.
Quality hoes.

Smith & Hawken Ltd., 2 Arbor Lane, Box 6900, Florence, KY 41022-6900; 1-800-776-3336.
Garden supplies.

Veldsma & Sons Inc., 160 Andrew Drive, Suite 100, Stockbridge, GA 30281; 1-800-458-7919.
Supplies for Christmas-tree growers and pumpkin marketers.

WoodMizer Inc., 8180 W. 10th Street, Indianapolis, IN 46214-2400; 1-800-553-0219, http://www.woodmizer.com.
Portable sawmills.

Worm's Way, 7850 N. Highway 37, Bloomington, IN 47404; 1-800-274-9676.
Garden supplies for urban farming.

Associations and Organizations

I am not listing very many associations, as there is one for every breed of animal and variety of crop under the sun. I have mostly picked generalized organizations that apply nationwide. For more information on associations and organizations in your area, talk to fellow farmers and your local extension office. Attend meetings and shows of local groups to increase your knowledge.

American Farmland Trust, 1920 North Street NW, Suite 400, Washington, DC 20036; 1-800-431-1499.
Organization dedicated to saving farmland.

American Herb Association, PO Box 1673, Nevada City, CA 95959; 916-265-9552.
Herb information.

American Livestock Breeds Conservancy, PO Box 477, Pittsboro, NC 27312; 919-542-5704.
Information on rare and endangered breeds of livestock.

The American Pastured Poultry Producers Association, 5207 70th Street, Chippewa Falls, WI 54729; 715-723-2262.
Information on pastured poultry.

The Biodynamic Farming and Gardening Association, PO Box 550, Kimberton, PA 19442; 610-935-7797, http://www.biodynamics.com.
Ecological farm management.

Center for Rural Affairs, PO Box 406, Walthill, NE 68067; 402-846-5428; http://www. cfra.org.
Information on national events affecting family farms.

Henry A. Wallace Institute for Alternative Agriculture Inc., 9200 Edmonston Road, Suite 117, Greenbelt, MD 20770-1551; 301-441-8777.
Information on sustainable farming systems.

The Herb Growing and Marketing Network, PO Box 245, Silver Spring, PA 17575; 717-393-3295; http://www.herbnet.com.
Herb information.

Herb Research Foundation, 1007 Pearl Street, Suite 200, Boulder, CO 80302; 303-449-2265.
Herb research information.

Kansas Rural Center, PO Box 133, Whiting, KS 66552; 913-873-3431.
Sustainable farming information.

Kerr Center for Sustainable Agriculture, PO Box 588, Poteau, OK 74953; 918-647-9123.
Information and packets on sustainable agriculture topics.

The Maine Organic Farmers and Gardeners Association, PO Box 2176, Augusta, ME 04338; 207-622-3118.
Organic information and factsheets.

Michael Fields Agricultural Institute, W2493 County Road ES, East Troy, WI 53210; 414-642-4028.
Sustainable farm promotion.

Michigan Integrated Food and Farming Systems, PO Box 4903, East Lansing, MI 48826.
Farmer support.

National Christmas Tree Association, 1000 Executive Parkway, Suite 220, St. Louis, MO 63141; 314-205-0944; http://www.Christree.org.
Represents both farmers and retailers of Christmas trees.

Northeast Organic Farming Association, 411 Sheldon Road, Barre, MA 01005; 508-355-2853.
Organic information.

Practical Farmers of Iowa, 2104 Agronomy Hall, Iowa State University, Ames, IA 50011.
Research for family farmers.

The Samuel Roberts Noble Foundation, PO Box 2180, Ardmore, OK 73502-2180; 580-223-5810, http://www.noble.org.
General farm information.

The Southern Sustainable Agriculture Working Group, PO Box 324, Elkins, AR 72727; 501-292-3714.

Sustainable Farming Association of Minnesota, Route 1, Box 4, Aldrich, MN 56434; 218-445-5475.
Sustainable farm information and newsletter.

Virginia Biological Farmers Association, Route 1, Box 46, Check, VA 24072; 540-651-4747.
Information on organic farming, newsletter.

Index

Note: Page numbers in *italic* indicate illustrations;
those in **boldface** indicate charts.

Other Storey Titles You Will Enjoy

Basic Butchering, by John J. Mettler Jr., D.V.M. Provides clear, concise, and step-by-step information for people who want to slaughter their own meat. 208 pages. Paperback. ISBN: 0-88266-391-7.

Fences for Pasture & Garden, by Gail Damerow. The complete guide to choosing, planning, and building today's best fences: wire, rail, electric, high-tension, temporary, woven, and snow. 160 pages. Paperback. ISBN 0-88266-753-X.

Growing and Selling Fresh-Cut Herbs, by Sandie Shores. Successful entrepreneur Sandie Shores offers the only guide for starting an herb business that includes complete growing and harvesting information with savvy advice for starting and maintaining a profitable company. Profiles of other successful entrepreneurs are included. 464 pages. Hardcover. ISBN 1-58017-128-1.

A Guide to Raising Beef Cattle, by Heather Smith Thomas. A bovine expert, Thomas explains facilities, breeding and genetics, calving, health care, and advice for marketing a cattle business. 352 pages. Paperback. ISBN 1-58017-037-4.

A Guide to Raising Chickens, by Gail Damerow. Expert advice on selecting breeds, caring for chicks, producing eggs, raising broilers, feeding, troubleshooting, and much more. 352 pages. Paperback. ISBN 0-88266-897-8.

A Guide to Raising Pigs, by Kelly Klober. Practical advice for buying, feeding, and caring for hogs, plus modern breeding and herd management. 320 pages. Paperback. ISBN 1-58017-011-0.

Keeping Livestock Healthy, by N. Bruce Haynes, D.V.M. A complete guide to preventing disease through good nutrition, proper housing, and appropriate care. 352 pages. Paperback. ISBN 0-88266-884-6.

Raising a Calf for Beef, by Phyllis Hobson. This no-nonsense how-to guide for beginners offers detailed information on choosing a calf, building and maintaining housing, nutrition, feeding, and daily care. Also includes instructions for slaughtering and butchering. 128 pages. Paperback. ISBN 0-88266-095-0.

Small-Scale Livestock Farming: A Grass-Based Approach for Health, Sustainability, and Profit, by Carol Ekarius. This natural, organic approach to livestock management produces healthier animals, reduces feed and health-care costs, and maximizes profit. Insights for working with nature instead of against it will help make your farm thrive. 224 pages. Paperback. ISBN 1-58017-162-1.

Successful Small-Scale Farming, by Karl Schwenke. Contains everything small-farm owners need to know, from buying land to organic growing methods and selling cash crops. 144 pages. Paperback. ISBN 0-88266-642-8.

These books and other Storey books are available at your bookstore, farm store, garden center, or directly from Storey Publishing, Schoolhouse Road, Pownal, Vermont 05261, or by calling 1-800-441-5700. Or visit our website at www.storey.com